The Oceans and Climate

GRANT R. BIGG

School of Environmental Sciences, University of East Anglia

CAMBRIDGE
UNIVERSITY PRESS

PUBLISHED BY THE PRESS SYNDICATE OF THE UNIVERSITY OF CAMBRIDGE
The Pitt Building, Trumpington Street, Cambridge CB2 1RP, United Kingdom

CAMBRIDGE UNIVERSITY PRESS
The Edinburgh Building, Cambridge CB2 2RU, UK http://www.cup.cam.ac.uk
40 West 20th Street, New York, NY 10011-4211, USA http://www.cup.org
10 Stamford Road, Oakleigh, Melbourne 3166, Australia

First published 1996
Reprinted 1998

Printed in the United Kingdom at the University Press, Cambridge

Typeset in Times 10/12 pt in Poltype™ [VN]

A catalogue record for this book is available from the British Library

Library of Congress cataloguing in publication data

Bigg, Grant R.
The oceans and climate / Grant R. Bigg
    p. cm.
Includes bibliographical references and index.
ISBN 0–521–45212–0 (hardcover)
1. Climatic changes.   2. Ocean–atmosphere interaction.   I. Title.
QC981.8.C5B54   1996
551.6′6–dc20   95–51194 CIP

ISBN 0 521 45212 0 hardback
ISBN 0 521 58268 7 paperback

Contents

Preface

In 1827 Jean-Baptiste Fourier, otherwise known for his contributions to mathematics, speculated that human activities had the capacity to affect the Earth's climate. In 1990 the International Panel on Climate Change produced a report detailing our current understanding of these activities, and speculated on what impact they might have on climate. In 160 years of great human endeavour much has been learnt but definitive evidence for climatic change driven by mankind remains elusive.

The oceans play a significant role in this tardiness of the climate system's response to our species. They store immense amounts of energy for months, decades or even centuries, depending on the region, depth and the nature of the interaction between the atmosphere and ocean. This storage capacity acts as a giant flywheel to the climate system, moderating change but prolonging it once change commences. The ocean also stores vast amounts of carbon dioxide.

In 1897 Svante Arrhenius discovered that the amount of carbon dioxide in the atmosphere affected the global temperature through the greenhouse effect. In 1938 G. S. Callendar showed that atmospheric carbon dioxide was increasing due to human activities. However, it has only been since the late 1960s that a rough estimate of the magnitude of the potential climatic effect has been possible. Even today the likely impact of a doubling of atmospheric carbon dioxide on raising global temperature is not known to within $3°C$; the global temperature at the height of the last Ice Age was only $4°C$ less than today.

A significant element in this uncertainty is the ocean. How is carbon dioxide and heat stored in the ocean? Are these mechanisms sensitive to climatic change? Could they interact with climatic change itself to accentuate, or lessen, such change? The exploration of these, among other, questions underlies this book.

The oceanic links to climate are complex and multi-faceted. The sciences of physics, chemistry and biology are interwoven in this tapestry. Therefore, after an introductory chapter on the climate system I devote chapters to the oceanic roles of each of these sciences, before examining some detailed ocean–atmosphere interactions affecting climate, and the role of the ocean in the past, and its potential role in the future climate.

My own introduction to this fascinating subject came through its physics, but I have aimed to make each science, and its links to the general problem of climate and air–sea interaction, understandable to readers coming from one of the other fields. English 'A' level standard physics, chemistry or mathematics would assist a reader but such a standard in only one of these subjects should not be a handicap. The book does not, therefore, contain many references – the climate literature is, in any case, vast and growing at an exponential rate – but does have a commented bibliography of the books and research papers that I have found most useful during its writing. This should provide the inquisitive reader with the tools to begin a more in-depth exploration of the subject. There is also a glossary of terms which are used repeatedly. The first use of each term is *italicized* in the main text.

The writing of such a book as this necessarily involves help from many sources. I would like to collectively thank the various publishers and authors who gave permission for diagrams to be used (individual identification is found in the appropriate figure legend). Internet has been an invaluable tool for tracking down data sets, and even for producing diagrams; the climate data site at Lamont-Doherty Geological Observatory merits particular thanks. I would also like to thank Fred Vine and Peter Liss for encouraging me to persevere with the book during its darkest days, and my editor, Conrad Guettler, for his keeping the literary ship on course. Phil Judge drew many of the diagrams and Sheila Davies photographed them. Most of all, my wife, Jane, put up with three years of writing angst and made the extremely valuable contribution of an arts graduate's criticism of the clarity of the science!

It is appropriate to end this preface with the following extracts from Shelley's *Ode to the West Wind* that encapsulate the tumultuous interaction between air and sea that this book explores:

> O wild West Wind, thou breath of Autumn's being,
> Thou, from whose unseen presence the leaves dead
> Are driven, like ghosts from an enchanter fleeing.
> ...
> Thou on whose streams, mid the steep sky's commotion,
> Loose clouds like earth's decaying leaves are shed,
> Shook from the tangled boughs of Heaven and Ocean,
> .... Thou
> For whose path the Atlantic's level powers
> Cleave themselves into chasms, while far below
> The sea-blooms and the oozy weeds which wear
> The sapless foliage of the ocean, know
> Thy voice, and suddenly grow gray with fear,
> And tremble and despoil themselves: oh hear!

1 The climate system

Traditionally climate has been defined as the average atmospheric state over at least a score of years, modulated by the seasonal cycle.

Such a definition conceals the temporal variability which produces the mean state, and the complexity of the underlying physical, chemical, biological, geographical, and astronomical processes contributing to the climate system. It also implicitly assumes that the climate of a locality does not change over decades, while accepting long-term changes such as glacial periods. In recent years this perspective of climate has changed. The strong coupling of different constituents of the climate system is now widely recognized, as is the fragility of 'stable' climate.

The traditional definition of climate contains two elements which lead us towards the concept of climatic change. First, the seasonal cycle demonstrates one scale of change through the direct impact of the annual change in solar radiation on the atmosphere and the *biosphere* (the collective term for plant and animal life). Contrastingly, the implication of stability over a number of years implies that basic balances exist within the system. It also implies that any decadal changes in the surface forcing of the atmosphere, from the land or ocean, are small; Chapters 5 and 6 will show that this is not always the case but in general this is a reasonable claim.

In recent years concern over potential climatic changes due to humanity's activities has arisen, although acknowledgement of this possibility was first made early in the nineteenth century. This book will explore the natural climate system, and potential changes, man-made (*anthropogenic*) or otherwise. Its dominant theme will be the contribution of the oceans to these processes. Observable changes to the climate due to anthropogenic inputs could be expected to have already occurred; Chapters 6 and 7 will show that recent climatic shifts are not, by the mid 1990s, categorically attributable to these inputs. The stability of the ocean's thermal response to change, its absorption of a significant proportion of anthropogenically derived compounds from the atmosphere, and its coupling with the atmosphere, form important pieces of the puzzle of climatic change.

To understand how the ocean affects, and is affected by, the climate we need to briefly consider the climate system as a whole. It is a complex, many-faceted system; Fig. 1.1 illustrates its major constituents and interactions. There are five components: the atmosphere, the ocean, the cryosphere (ice sheets and sea ice), the biosphere, and the geosphere (the solid earth). The system is driven by short wavelength, visible and ultra-violet, solar

1

Fig. 1.1. A schematic
diagram of the climate
system. [From Bigg,
1992d]

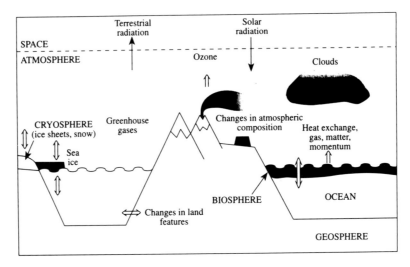

radiation, with longer wavelength, infra-red, radiation being emitted into space to maintain thermal equilibrium. In the rest of this chapter we will briefly examine the different components of the climate system, including the timescales on which it operates. However, we will begin with a short discussion of the basic energy source driving the climate – radiation from the Sun.

1.1   Solar radiation

The interior of the Sun, where the nuclear reactions occur that ultimately lead to life on Earth, is incredibly hot, at a temperature of several million degrees Celsius. However, the *electromagnetic radiation* (see Appendix A) that provides the energy for the climate system is derived from the outer layers of the Sun. The greatest amount of radiation comes from the *photosphere*, a layer some 300km thick in the solar atmosphere. This varies in temperature from 10 000K[1] at the bottom to 5000K at the top. Outside the photosphere are much less dense regions – the chromosphere and corona. While these outer regions are at much higher temperatures, up to millions of degrees in the corona, their low density means that they radiate relatively little energy. Most of this is at very short, X-ray and gamma-ray wavelengths which affect the upper atmospheres of the planets (see §3.7) but do not penetrate into the lower atmosphere.

The Sun appears to us as (almost) a *black body*. That is, the *spectrum* and total energy of electromagnetic radiation emitted from the Sun (as from all surfaces, and indeed molecules) is a function of its temperature. The total energy flux, E, emitted by a black body follows the *Stefan–Boltzmann Law*:

$$E = \sigma T^4 \tag{1.1}$$

where σ is the Stefan–Boltzmann constant and T is the temperature in

---

[1] The absolute scale of temperature is in degrees Kelvin. In this scale 0K is the coldest possible temperature when all molecular motion has stopped. The freezing point of water, 0°C, is 273.16K in this scale. Note, however, that a change of 1K is equivalent to a change of 1°C.

Fig. 1.2. The Sun's
spectrum, seen from space
(dashed line). Both scales
are logarithmic. For
comparison, a Planck
spectrum for a temperature
of 5785K is shown (solid).
Note the accentuation of
long and short wavelength
energies emitted by the
Sun, particularly during
solar flares.

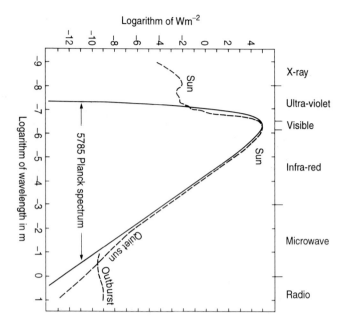

degrees Kelvin (a list of constants and their values can be found in Appendix
A). The energy density, E_λ, or radiant energy per unit wavelength, λ, per unit
volume per second, is given by

$$E_\lambda = \frac{8\pi c}{\lambda^5}\left[\frac{1}{e^{hc/\lambda kT} - 1}\right] \tag{1.2}$$

where c is the speed of light, k is Boltzman's constant and $\hbar$ is Planck's
constant. The Sun's spectra, as observed from space (Fig. 1.2), obeys (1.2) for
a temperature near 6000K. However, for very small (X-ray) and very long
(microwave) wavelengths the solar spectrum is enhanced due to contributions from the outer regions of the solar atmosphere (see §§3.7 and 7.1.1).

The vast majority of the energy which reaches the Earth comes from the
ultra-violet through visible to infra-red part of the spectrum. The peak
energy is in the visible, near wavelengths that we see as the colour blue. The
amount of energy emitted by the Sun is probably nearly invariant on the
time scales we are considering in this book. At the Earth's distance from the
Sun this is the *solar constant* of 1.38 kWm^{-2}. On very long time scales,
comparable with the life of the planet, astrophysicists believe that the Sun's
irradiance varies dramatically as the supply of fuel within the Sun changes.
We will see in §1.8 that variation in the Earth's orbit can affect the amount
of energy by a few per cent, on time scales of thousands of years. However,
over several decades to centuries solar irradiance is thought to vary by
significantly less than this. Satellite measurements extend back only to 1978
and these reveal irradiance changes of only 0.1% between *sunspot* maxima
(higher) and minima (lower). This does not, however, preclude larger
changes in more active beats of the 11 year solar cycle, or the existence of
frequencies in the Sun's behaviour of which we are unaware (see §7.1.1).

1.1.1 *The effective temperature of the Earth*

If the Earth was a sterile planet like the Moon, with no atmosphere, oceans or biosphere what temperature would we expect the surface to possess, given the solar constant, S, at the Earth's astronomical position? If we think of the Earth as a flat disc, viewed from the Sun, then the surface area illuminated by solar radiation is πr^2, where r is the radius of the Earth. The energy absorbed is thus $(1 - a)S\pi r^2$, where a is the *albedo*, or the amount of energy reflected from the Earth back into space ($\sim 30\%$).[2] For equilibrium between the absorbed solar radiation and the emitted radiation from the whole Earth's surface of area $4\pi r^2$, the Earth's temperature, T_E, will therefore, from (1.1), be

$$T_E = \left[\frac{(1 - a)S}{4\sigma} \right]^{0.25} \tag{1.3}$$

Equation (1.3) gives a surface temperature for this hypothetical atmosphere-less planet of 255K, or $-18°C$, much colder than the Earth's average surface temperature of about $16°C$. This effective planetary temperature is more typical of the real atmospheric temperature at a height of about 6km above the surface. The atmosphere clearly has a significant impact on the distribution of the energy contributing to this effective temperature and will thus be the first component of the climate system to be considered.

1.2 The atmosphere

The atmosphere is a largely homogeneous mixture of gases, both horizontally and vertically, over the height range important for climate: namely the *troposphere* and *stratosphere* (Fig. 1.3). The composition of this apparently stable mixture, air, is shown in Table 1.1. The balance of the dominant constituents of air is thought to have evolved considerably over the lifetime of the planet, for instance, oxygen is likely to have been a product, rather than a necessity, of life (the *Gaia hypothesis*). The climate system's immense natural variability will be a recurring theme of our discussion throughout the book.

The temperature of the atmosphere varies strongly both in the vertical and with latitude. The latter is due to an imbalance in the radiation received over the Earth's surface throughout the year (Fig. 1.4). The circulation of both the atmosphere and ocean are ultimately derived from this energy imbalance; they act to counter it, in the ratio of about 3:2 respectively.

The vertical temperature distribution shown in Fig. 1.3 comes about because the atmosphere is basically heated from two sources: the ground and the upper stratosphere (although we will see in the next sub-section that this is a significant simplification). The ground (or ocean surface) is a heat source since some 46% of the incoming solar radiation is absorbed there. There is also an important heat source between 30 and 50 km above the

---

[2] Note that this average planetary albedo assumes that the hypothetical sterile Earth has the same net reflectivity as the real Earth–atmosphere system. Thus this albedo is *not* the surface reflectivity (see §1.4).

Table 1.1. *The major constituents of the atmosphere*

Gaseous constituent	Molecular form	Proportion (%)
Nitrogen	N_2	78.1
Oxygen	O_2	20.9
Argon	Ar	0.93
Water vapour	H_2O	variable: 0.1–1
Carbon dioxide	CO_2	0.0355
Methane	CH_4	0.000172
Nitric oxide	N_2O	0.000031
Ozone	O_3	variable $\sim$ 0.000005

Fig. 1.3. The zonal mean vertical profile of temperature during June at 45°N.

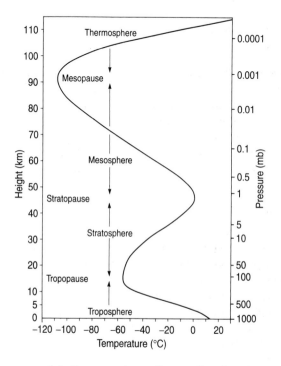

ground, in the *ozone layer*. Ozone, O_3, absorbs a photon of ultra-violet light – denoted v because this is the energy of a photon of frequency v – dissociating in the process to form an oxygen molecule and an energetic oxygen atom, O^1D, where one electron is displaced into a higher energy state than in the ordinary oxygen atom (see Appendix B for the Periodic Table of the Elements and a discussion of electron orbitals). This can then react with an oxygen molecule to reform ozone as part of the Chapman cycle:

$$O_3 + v \rightleftharpoons O_2 + O^1D \qquad O^1D + O_2 + M \rightleftharpoons O_3 + M \qquad (1.4)$$

The air molecule, M (that is, predominantly N_2 or O_2), is necessary in the second reaction in (1.4) as the reaction produces excess energy. This is carried away by M thereby stabilising O_3, which would otherwise dissoci-

Fig. 1.4. Contour graph of
the daily average insolation
at the top of the
atmosphere as a function
of season and latitude. The
contour interval is 50
Wm^{-2}. The heavy dashed
line indicates the latitude of
the sub-solar point at
noon. [Fig. 2.6 of
Hartmann (1994), *Global
physical climatology*.
Reprinted with permission
from Academic Press.]

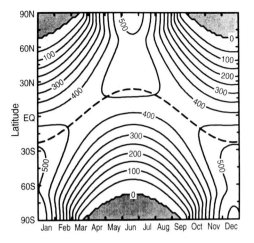

Fig. 1.4. Contour graph of the daily average insolation at the top of the atmosphere as a function of season and latitude. The contour interval is 50 Wm^{-2}. The heavy dashed line indicates the latitude of the sub-solar point at noon. [Fig. 2.6 of Hartmann (1994), *Global physical climatology*. Reprinted with permission from Academic Press.]

ate. Such a reaction is called *exothermic*. Hence the air becomes warmer, as the temperature of a medium is merely a reflection of the average kinetic energy of its molecules.

The reactions in equation (1.4) are only part of the full Chapman cycle, which also contains reactions involving photo-dissociation of O_2, and reactions between the excited oxygen atoms themselves and O_2. Some of these are likewise exothermic, adding to the energy which is transferred, via chemical reaction, from solar radiation to the middle atmosphere. There are many other reactions involving ozone, some of which will be discussed in §7.2.1.

The lower atmosphere is therefore heated both from above and below. Between these regions is a zone, in the lower stratosphere, where the energy from these heated regions only weakly penetrates. This is strongly stratified, which means that there are large vertical gradients in the concentrations of trace constituents of the air and the *potential temperature* (see Appendix C). The tropopause, at the bottom of the stratosphere where the gradients are greatest, resists penetration by cloud convection, or even diffusion. The well-mixed region below this, the troposphere, is the part of the atmosphere that we will be largely concerned with, because of its direct interaction with the oceans.

The strong heating of the surface at the equator (Fig. 1.4) makes the air less dense, forcing it to rise. Air flows towards this region of rising air, which is concentrated in a narrow band around the globe known as the Inter-Tropical Convergence Zone, or ITCZ. Aloft, the rising air moves poleward to compensate for the surface flow. In the late seventeenth century, when Halley first proposed this mechanism for driving the atmospheric circulation (modified 50 years later by Hadley) it was believed that this *Hadley cell* extended to the polar regions. This seemed logical, as polar air is cold, and so relatively dense, and should therefore flow towards the low pressure regions of the tropics in order to transfer heat from the equator to the poles and so maintain the Earth's thermal equilibrium. By the nineteenth century this idea was seen to be too simplistic. The Coriolis force, due to the solid Earth and moving atmosphere revolving at slightly

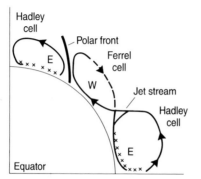

Fig. 1.5. A schematic cross-section of the zonal mean circulation in the troposphere. The dotted upper region of the Ferrel cell indicates that it is a less constant feature. The crosses (and E) show where the surface flow has an easterly component, while the W shows where the surface flow is predominantly westerly.

different rates (see §2.5.2), gives this converging near-surface wind a westward component, resulting in the observed easterly[3] Trade winds. Ferrel therefore proposed an intermediate 'Ferrel cell' in mid-latitudes. Modern observations support this, as shown schematically in Fig. 1.5, a zonal cross-section of the tropospheric flow.

The general circulation of the lower troposphere is shown in Fig. 1.6, and the sea level pressure field for northern winter in Fig. 1.7. The ascending air of the equatorial region is shown by the low pressure. To replace this, easterly winds flow equatorwards driven by high pressure in the sub-tropics, where the air in the tropical Hadley cell descends (Fig. 1.5). This latter, relatively calm, zone has strong westerly winds on its poleward side, which, in turn, lie equatorward of another region of low pressure near 60° of latitude. This region of sub-polar low pressure forms the ascending branch of the polar Hadley cell of Fig. 1.5, with easterly winds at the surface due to polar high pressure. The most vigorous part of this system is where the tropical Hadley cell meets the mid-latitude Ferrel cell. Here the upper level convergence of air produces an extremely strong westerly *jet stream* in the upper troposphere (Fig. 1.5). This often has a secondary maximum over the mid-latitudes, above the polar front (Fig. 1.5). This polar front jetstream steers the transient pressure systems that we experience on the ground in the mid-latitudes. The latter systems are a significant mechanism in the redistribution of heat from equator to pole.

The mainly zonally symmetric structure of the general circulation is mostly due to the latitudinal distribution of the solar radiation received by the Earth; the distribution of land and sea over the Earth's surface distorts the zonality. Some aspects of this latter interaction will be discussed in Chapters 2 and 5.

1.2.1 The greenhouse effect

The vertical profile in Fig. 1.3 shows a decline in temperature of about 6.5°C per kilometre in the troposphere. It can be shown that ascending 'dry' air,

---

[3] Confusingly, meteorologists and oceanographers follow different conventions when specifying the direction of fluid flow. Meteorologists use the direction from which the wind has come to describe it, while oceanographers take the direction in which the flow is going. Thus, an east*erly* wind to a meteorologist is a west*ward* wind to an oceanographer! This unfortunate difference is too entrenched to be easily altered, and this book will use the convention appropriate to the fluid medium being described.

Fig. 1.6. The mean surface
wind field in (a) January,
and (b) July. The data is
from Oort (1983).

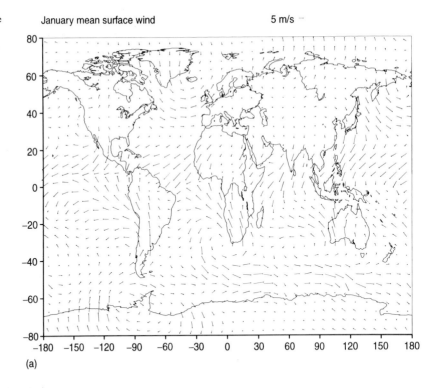

(a)

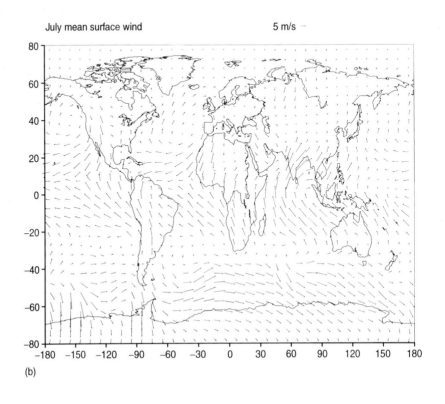

(b)

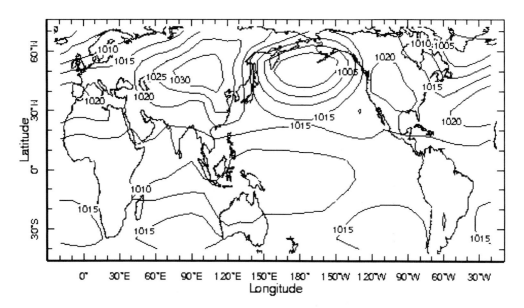

Fig. 1.7. The mean surface pressure field in January. The contours are every 5 mb. The data is a mean of 11 years of National Meteorological Center model analysis fields.

i.e. air without clouds, changes temperature because of expansional cooling by 9.8°C for each kilometre in *adiabatic* vertical motion (the latter occurs if a parcel of air does not exchange any heat with its surroundings, as is a good approximation in, for example, the formation of *cumulus* clouds – see Appendix C). Within a cloud the decline of temperature with height in vertical motion can approach the typical value of Fig. 1.3, due to the release of *latent heat* upon condensation of water vapour. However, substantially less than half of the troposphere contains cloud at any one time so other processes must be lowering the *environmental lapse rate.* Diffusion and advection of heat from the stratosphere, the ground, or surrounding air masses is partially responsible but the major reason for the enhancement of tropospheric temperatures is the *greenhouse effect.*

A number of low concentration, or *trace*, gases in the atmosphere are unresponsive to illumination by short wavelength radiation from the Sun but absorb energy of infra-red wavelengths. The gas molecules do this by increasing their *vibrational* and *rotational* energies, rather than their kinetic energy. How this happens can be illustrated by the water molecule, shown in Fig. 1.8. The bond angle between the hydrogen atoms of an ordinary water molecule is 105°, but if a *photon* of a certain wavelength of infra-red radiation (6.27 μm)[4] collides with the molecule the energy of the photon can be converted into vibration of the hydrogen bonds, such that the angle between the hydrogen atoms undergoes rapid oscillation of a few degrees. Other forms of oscillation can be excited by wavelengths of 2.66 or 2.74 μm. The *absorption spectra* of H_2O, shown in Fig. 1.9, is more complex than just these three wavelengths, however, as multiples, or *harmonics*, of the principal absorption wavelengths can also be absorbed. In addition wavelengths which are sums, or differences, of these three (and their

[4] 1 μm = 10^{-6} m.

Fig. 1.8. A schematic
diagram of a water
molecule. H represents a
hydrogen atom and the
central O an oxygen atom.
The solid lines show the
bond positions.

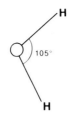

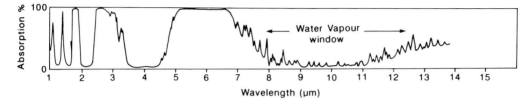

Fig. 1.9. The absorption
spectrum of water vapour.
Note the region 8–12 μm,
known as the 'water
vapour window' where
there is little absorption of
infra-red radiation by the
water vapour molecule.

harmonics) also show a degree of absorption, although generally of much less intensity.

Equation (1.2) shows that the wavelength of electromagnetic radiation emitted by an object is inversely related to its temperature. Thus the mean wavelength of the radiation emitted by the Earth's surface, and within the atmosphere itself, will be longer than that of the incoming radiation from the Sun, as the latter has a surface temperature of about 6000K compared to a typical Earth surface temperature of 289K. Fig. 1.10 depicts a typical energy spectrum, seen from the tropopause, of the radiation from the Earth's surface, with the absorption by trace gases shown by shading. There are regions of the spectrum, such as wavelengths shorter than 8 μm and from 15–20 μm, where the infra-red radiation is almost totally absorbed by atmospheric gases. It is this absorption, and the associated re-emission of energy, much of which warms the troposphere, that is called the greenhouse effect. This name is a misnomer as the physical mechanism involved in keeping a greenhouse warm is totally different to this radiative physics. There is a small contribution from glass being transparent to solar radiation, but partially reflective to the out-going infra-red radiation from the air and soil within the greenhouse. However, greenhouses are warm predominantly because the enclosed space eliminates convection, and hence mixing with cooler air.

The principal greenhouse gases, and their relative contribution to the greenhouse effect, are shown in Table 1.2. The percentages shown are not strictly additive because the absorption ranges of the different gases overlap. Table 1.2 also gives the fundamental absorption wavelengths of these molecules, but the complexity of the absorption spectra, with their harmonics and linear combinations of these fundamental wavelengths, must be remembered (see Fig. 1.9). Water vapour is two to three times as important in the total greenhouse effect as carbon dioxide. This fact is often neglected in discussions of greenhouse warming because water vapour is highly variable in concentration, both in space and time, making it difficult to isolate its global effect. It will, however, be vital to much of our later discussion.

Table 1.2. *The greenhouse gases*

Gas	Basic absorption wavelengths (μm)	Contribution
Water vapour (H_2O)	2.66, 2.74, 6.27	55–70%
Carbon dioxide (CO_2)	4.26, 7.52, 14.99	25%
Methane (CH_4)	3.43, 6.85, 7.27	5%
Nitrous oxide (N_2O)	4.50, 7.78, 16.98	2%
Chlorofluorocarbons (CFCs)	typical bonds: 9.52, 13.8, 15.4	1%
Ozone (O_3), sulphur dioxide (SO_2), other oxides of nitrogen, carbon monoxide (CO), etc.		<1% each

Fig. 1.10. Earth's surface radiation spectrum, seen at the tropopause. The dotted line is the black body emission for a typical surface temperature of 294K (21°C). The solid line is the observed spectrum, with the shaded region between denoting the energy absorbed by gases in the troposphere.

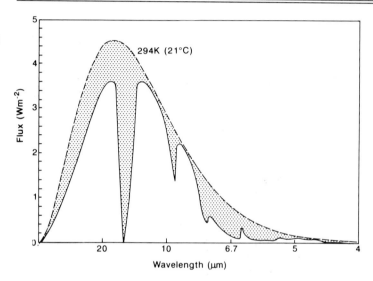

The importance of the greenhouse effect to the *heat budget* of the atmosphere, and therefore the climate system, is shown in Fig. 1.11. The *feedback* mechanism between the radiation from the Earth's surface and the greenhouse re-radiation raises the amount of energy available to heat the surface from 70% of the incident solar radiation (in the absence of the atmosphere) to 133%. Present and future changes to the amounts, and proportions, of the trace gases that contribute to this effect may change these figures and so have implications for the global climate. A number of the greenhouse gases have increased significantly in concentration in the last 200 years. This may be linked to the rise in global average surface temperature of about 0.5°C over the past century. Detailed discussion of these variations will be delayed to Chapters 6 and 7, which examine natural and anthropogenic alterations to the climate system.

1.2.2 Reflected radiation

Another major pathway for energy in the atmosphere is reflection from the surface, clouds or airborne particles. About 30% of the incident radiation is

Fig. 1.11. Global average
pathways for energy in the
atmosphere. A notional 100
units comes from the Sun.
[From Bigg, 1992a]

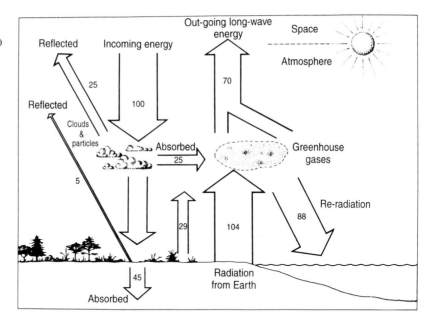

Fig. 1.11. Global average pathways for energy in the atmosphere. A notional 100 units comes from the Sun. [From Bigg, 1992a]

so lost from the climate system. The surface accounts for a sixth of this (the surface albedo) but the predominant loss is from the atmosphere. Variation in the *cloud amount* and type, the amount of suspended volcanic debris or solid chemical aggregates, and the characteristics of the Earth's surface can change the magnitude of this energy sink. While most of these processes will not be directly affected by the ocean we will see several exceptions later.

1.3   The oceans

The oceans cover 361 million square kilometres, or 71% of the surface area of the globe, almost two and a half times the land area. To the surface observer this immense area seems almost featureless compared to the land, with only icebergs and waves to give a vertical dimension. However, beneath the water surface the ocean floor shows all the orographic richness of the land.

Fig. 1.12 shows the percentage of the Earth's surface in different height bands, relative to mean sea level. The first striking feature is the greater average depth of the ocean compared with the land's altitude. Much of the oceans are more than 3000 m deep, while little of the land surface is above 3000 m in altitude. This discrepancy also appears in the extremes of orography; Mount Everest is 8848 m in altitude, but the deepest point in the Marianas Trench, east of the Phillipines, is 11022 m below sea level. Large areas of the ocean have relatively little variation in depth; these regions are known as abyssal plains. They are usually deeper than 3000 m.

Separating these deep, oceanic plateaus is more 'mountainous' *bathymetry*. For example, the Mid-Atlantic Ridge begins north of Eurasia and essentially splits the Atlantic Ocean into two halves. This ridge then continues eastward across the Southern Ocean, curving northward into the

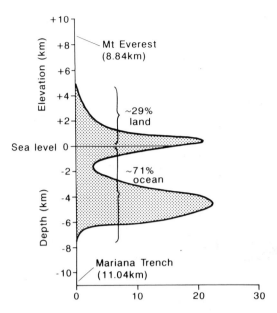

Fig. 1.12. Variation of the elevation of the Earth's surface. The mean elevation of the land is 0.84 km, while the mean depth of the ocean is 3.8 km.

east Pacific as the East Pacific Rise. A spur extends northwards from this main ridge into the western Indian Ocean.

The character of the oceanic perimeters varies considerably. In places the continental shelves adjacent to the coast are hundreds of kilometres across, such as off part of western Europe and northern Australia. Elsewhere the shelf may be only a few tens of kilometres wide; a good example is off the west coast of South America. The continental slope, which joins the shallow coastal zone to the deep ocean, can be very steep in such regions, with average gradients greater than 1 in 10. In §1.3.2 we will see that these various bathymetric structures are important in guiding the oceanic circulation, particularly the deep water flow.

1.3.1 Chemical composition of the oceans

The oceans are 96.5%, by volume, water. The water molecule has properties that are important for the movement of heat, momentum and climatically active gases (including water itself) between the atmosphere and the oceans. The remaining 3.5% of the oceanic solution – dissolved salts, particles, organic material and gases – plays an inordinately large role in such climatic processes and the circulation in the ocean. We will here consider some chemical properties of water and the dissolved salts; more details of the chemical and biological processes in the ocean which contribute to the climate system will be the subject of Chapters 3 and 4.

Water, or H_2O, is a very special molecule. Other compounds with similar molecular weight, such as methane, CH_4, or ammonia, NH_3, are gases at room temperature. Water, by contrast, is a liquid, and is readily found in its solid state (ice) below 0°C. The reason for this unusual behaviour lies in the molecular structure, already seen in Fig. 1.8. A water molecule is composed of an oxygen atom, bonded to two hydrogen atoms separated by an angle of

Table 1.3. *Some physical properties of liquid water*

Latent heat of fusion	3.33×10^5 Jkg^{-1}K^{-1}
Latent heat of vaporization	2.25×10^6 Jkg^{-1}K^{-1}
Specific heat	4.18×10^3 Jkg^{-1}K^{-1}
Surface tension	7.2×10^9 Nm^{-1}
Maximum density	1.00×10^3 kgm^{-3}
Heat conductivity (at 290 K)	5.92×10^{-2} Js^{1}m^{-1}K^{-1}
Molecular viscosity (at 293 K)	1.0×10^{-2} Nsm^{-2}

$105°$. As a result of the re-arrangement of the atomic electron orbitals (see Appendix B), the oxygen atom accrues a small net negative charge, while the hydrogen atoms gain a small positive charge. This makes the molecule dipolar and allows groups of molecules to form aggregates, with oppositely charged portions of the molecules adjoining each other, held together by *coulombic* forces. These structures resist break-up and permit water to be in a less energetic state than other, similar molecular weight, compounds at ordinary temperatures. Such unusual bonding leads to a number of important physical properties. These are summarised in Table 1.3, but are worth elaboration because we shall see their effects in later chapters.

The latent heat of fusion (the energy required to melt 1 kg of ice), and the latent heat of vapourization (that needed to evaporate 1 kg of water), are among the highest for any substance. This has important implications for the climate system as, conversely, this energy is released to the environment when the water changes state back to a more ordered structure. For instance, when water vapour condenses to form water droplets in a cloud, the energy latent within the vapour is released as heat and contributes to the driving energy of the cloud-producing process.

Related to these properties is the specific heat, highest of all solids and liquids except ammonia. This is the amount of energy required to increase the temperature of one kilogramme of the substance by 1°C. Dry air requires less than a quarter of the energy water needs to heat a kilogramme by 1°C, and when the thousand-fold difference in density is taken into account, it is quickly seen that the ocean will be much slower in responding to heating, or cooling, than the atmosphere. This is an extremely important property climatically, as not only does it explain the smaller annual range in temperature of maritime climates, but it also points to the ocean's ability to act as a flywheel for longer term climatic change. Energy can be both stored and released over decades, or even centuries, by the ocean while the atmosphere reacts to energy changes with time delays of only a few weeks.

The heat conductivity and molecular viscosity of water are also strongly affected by the inter-molecular forces, being unusually high and low respectively. These parameters give the mixing ability of the liquid with respect to heat and molecular motion. However, mixing within water principally occurs because of stirring by eddies within the fluid, which are on a much bigger scale than the molecular processes for which conductivity and viscosity are appropriate. Therefore these properties are not significant for the processes with which we are concerned. A final physical property of note that derives from liquid water's unique structure is the high surface

Table 1.4. *Concentration of major ions in sea water*

Constituent	Ion	Average concentration in sea water of salinity 35 psu	Average concentration in river water (part per thousand)
Chloride	Cl^-	19.350	0.0078
Sodium	Na^+	10.760	0.0063
Sulphate	SO_4^{2-}	2.712	0.0012
Magnesium	Mg^{2+}	1.294	0.0041
Calcium	Ca^{2+}	0.412	0.0150
Potassium	K^+	0.399	0.0023
Bicarbonate/ carbonate	HCO_3^- / CO_3^{2-}	0.145	0.0588

tension. This is related to the force needed to break the air–water interface: a high value is detrimental to the speed of gas and particle exchanges between the air and water. This will be of importance in later chapters.

Another consequence of the molecular structure of water is its dissolving power. Sea water is a mixture of many compounds; the main ingredients, apart from water itself, are shown in Table 1.4. The addition of these salts has its own effect on the properties of the mixture. The freezing point of sea water is about $-1.8°C$, rather than $0°C$. This lowering of the freezing point of water upon the addition of salts underlies the salting of roads in winter when near freezing temperatures are expected. The density is also affected by the addition of salt. A typical surface sea water density is about 1026 kgm^{-3}, an increase of 2.6% above that for pure water (see Table 1.3; note that the density of air near sea level is only about $1.2\,kgm^{-3}$). The density of sea water is a complicated function of temperature, salinity (the concentration of dissolved salts)[5] and pressure (see, for example, Gill 1982, Appendix 3). However, because of the strength of the inter-molecular forces within water near its freezing point, it is found that salinity has most effect on density at low temperatures, while temperature exerts the predominant influence at higher temperatures.

The main input of particulate or dissolved material to the oceans occurs through riverine input; there is a small contribution from wind-blown (aeolian) deposits and precipitation. The globally averaged riverine chemical composition is distinctly different from the sea water composition shown in Table 1.4. Bicarbonate (HCO_3^-) is the dominant riverine *anion* and calcium (Ca^{2+}) the most prevalent *cation*. Neither sodium (Na^+) nor chloride (Cl^-) contribute large percentages to the total dissolved ion concentration.

Sea water appears to be of remarkably stable composition; the salinity may vary but the proportions of the different salts remain almost constant. Therefore, for a considerable time, perhaps hundreds of millions of years, the riverine input has been in balance with processes which remove the

---

[5] Since 1982 a salinity scale based on the electrical conductivity of sea water has been used. The average salinity of the oceans is 35×10^{-3}, or 35 practical salinity units (psu) in this scale (a dimensionless number). Salinity values in psu are essentially identical to a measure of parts per thousand by weight.

salts, such as sedimentation and ejection into the atmosphere. The excess chloride in sea water, in comparison with the riverine source, is thought to have come from volcanism early in the Earth's history. Volcanic eruptions would have emitted large quantities of the very soluble gas HCl. Dissolution of this gas in the sea forms a very weak hydrochloric acid solution. The input of bicarbonate over time has neutralised this, to leave sea water as slightly alkaline (pH $\sim$ 8.0). Another non-riverine input to the oceans which has contributed to the concentration of trace constituents is submarine hydrothermal activity on mid-ocean ridges. These sites are sources for dissolved gases like helium, and some metals, for example manganese.

The chloride ion concentration in sea water may be explained by early volcanism, but sodium is not a large component of volcanic gases. Sodium must therefore attain its abundance by other means. This leads us to an important concept in the chemistry of the environment, namely the concept of elemental *cycling*. Many chemical elements cycle repeatedly through various parts of the Earth's outer crust, atmosphere and oceans in such a way that the concentration in each component of the system is stable over long periods of time. The carbon cycle is the best known of these, and will be considered in detail in Chapter 3. Other environmentally important cycles include sulphur, calcium, sodium and lead. The time taken for an atom to complete one cycle is determined by the average *residence time* of an atom of the element concerned within the different components of the cycle. In the ocean, for instance, this is defined as the total amount of the element in the ocean divided by the riverine and aeolian input per year. This definition implicitly assumes the cycle is in long-term balance.

To explain the abundance of sodium, relative to calcium, in sea water we can use the concept of residence time. Calcium has a much shorter residence time in the oceans than sodium. This is because it is a major constituent in the skeletons of marine organisms and is therefore easily lost to the ocean through settling of dead organisms onto the sea floor. It is therefore taken out of the oceans into the sediments sufficiently fast for more sodium, from riverine inputs, to accumulate in the oceans than calcium.

Cycling can also be applied to important molecules, as well as elements. The hydrological cycle is merely the cycle of water through the climate system. This is illustrated in Fig. 1.13. The residence time of water in the ocean is 3220 years. This can be compared with the time taken for material injected into the deep ocean to thoroughly mix around the globe (typically 1000 years). By contrast, the residence time for water vapour in the atmosphere is only 10 days and the mixing time is of the order of a month.

1.3.2   Ocean circulation

A schematic of the global ocean circulation is shown in Fig. 1.14. This is known as the *thermohaline* circulation because it is driven by density contrasts. The basic structure consists of deep water being carried towards the Pacific Ocean, which then upwells towards the surface and is transported back to the downwelling regions in the North Atlantic (principally

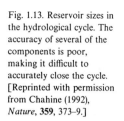

Fig. 1.13. Reservoir sizes in the hydrological cycle. The accuracy of several of the components is poor, making it difficult to accurately close the cycle. [Reprinted with permission from Chahine (1992), *Nature*, **359**, 373–9.]

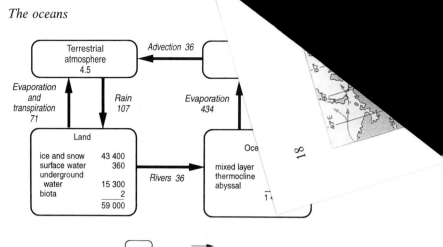

Reservoirs in 10^{15} kg   Fluxes in 10^{15} kg yr^{-1}

Fig. 1.14. Schematic of the thermohaline circulation of the global ocean – the Conveyor Belt. The broken arrows represent the major surface components of the circulation. The continuous line denotes the deep water circulation, emanating from source regions denoted by open circles. Slow upwelling in the Indian and Pacific Oceans closes the circuit.

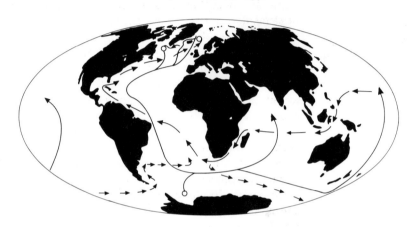

the Norwegian–Greenland Sea and the Labrador Sea) and the Weddell Sea. This cycle, known as the *conveyor belt*, is very important in the climate system. We will see in later chapters that it provides a stabilising effect on climate, because of its long time scale (see the last section), but can also cause abrupt climatic change in the space of a few decades if it is disturbed in certain ways.

The conveyor belt mechanism is naturally a gross simplification. The mean surface circulation is shown in Fig. 1.15. It has several shared features in each basin. Sub-tropical *gyres* rotate *anticyclonically* in each of the main ocean basins. The western margins of each of these gyres have strong poleward currents, such as the Gulf Stream in the North Atlantic. Poleward of these gyres there is some evidence in the northern hemisphere for *cyclonic* sub-polar gyres, where the westerly winds change to polar easterlies. In the southern hemisphere the water is able to flow around the entire globe, driven by the strong westerly winds at these latitudes. Sub-polar gyres exist in the Weddell and Ross Seas, poleward of this Antarctic Circumpolar Current.

The surface flow in the tropics consists of strong westward flowing currents at, and near, the equator. These are extensions of the tropical arm

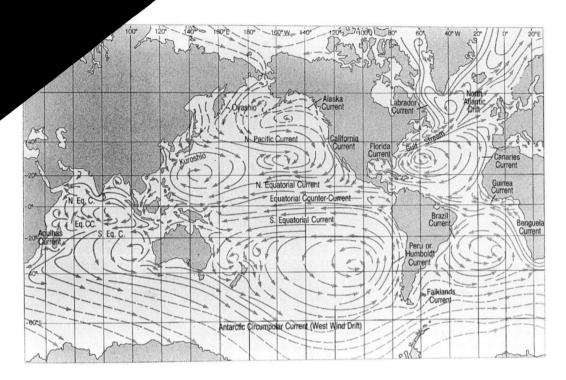

Fig. 1.15. The global
surface current system.
Cool currents are shown
by dashed arrows; warm
currents are shown by solid
arrows. The map shows
average conditions for
winter months in the
Northern Hemisphere;
there are local differences
in the summer, particularly
in regions affected by
monsoonal circulations.
[Fig. 3.1 of Open
University Course Team,
1989. Reprinted with
permission from
Butterworth–Heinemann.]

of the sub-tropical gyres. Between these two westward currents a counter-current, flowing eastwards, is usually found. There is also a strong eastward current below the surface on the equator, the equatorial under-current. The equatorial currents are intimately coupled with the atmosphere and will be discussed further in Chapters 2 and 5.

The deep circulation, shown in Fig. 1.16, conveys water that has sunk in the polar regions throughout the world oceans. In the Greenland and Norwegian Seas during winter the surface waters are strongly cooled making them denser. This dense water then overturns, probably in very localised regions up to a few tens of kilometres in diameter. In the southern hemisphere, particularly in the Weddell Sea, ice formation leaves a greater concentration of salt in the water beneath, as salt tends to be expelled from the ice lattice as it forms. This dense water also sinks. These two distinct types of water, or *water masses*, then travel equatorwards from the polar regions, to form the deep waters of the world's oceans. This deep water circulation is driven by subtle differences in temperature and salinity.

The water at intermediate depths also comes from the sinking of water masses, but those formed in less extreme circumstances. One such import-ant contribution comes from the Mediterranean Sea. Intense evaporation raises the salinity of this basin above that of the North Atlantic. This dense, saline water sinks to form the deep water of the Mediterranean basin. At the Strait of Gibraltar surface water flows into the basin from the North Atlantic, to compensate for the evaporation. To conserve mass locally, the deep water flows out into the Atlantic beneath. This warm, salty water is evident in the intermediate layers of the North Atlantic for thousands of

Fig. 1.16. The deep
circulation of the global
ocean. The two main
sources of deep water are
shown by open circles. The
deep water originating
from the North Atlantic is
slightly less dense than the
deep water of Antarctic
origin; hence its path is
shown by dashed arrows
until it merges with the
latter (whose path is shown
by a continuous line).

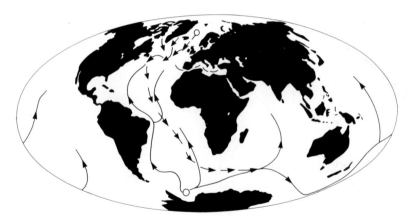

kilometres, contributing about 6% of the North Atlantic's salinity. It also acts to pre-condition the water entering the Norwegian–Greenland Sea, that will later overturn and produce deep water, by making it denser than it would otherwise be.

1.4   The cryosphere

Ice covers about 5.7% of the Earth's surface and contains 2.05% of the Earth's supply of water (the oceans contain 97.25%). It is very variable seasonally and comes in many forms: continental ice sheets, mountain glaciers, shoaled ice shelves, sea-ice, snow and perma-frost.

In this book we will mainly be concerned with sea-ice. However, land-ice has a high albedo – 0.95 for freshly fallen snow, over 0.4 for old snow and ice – and low temperature. As the average albedo of the Earth is 0.3, and of the oceans 0.08, the ice cover both on land and on the oceans drastically reduces the heat energy entering the climate system (see §2.1 for albedos of other substances). Ice cover also tends to reduce the input of moisture, and hence latent heat, to the atmosphere by evaporation. The quantity of global ice has varied significantly in the past. A hundred million years ago there was probably almost none. As recently as 15 000 years before present (BP) continental ice covered much of Canada, the northern United States, northern Siberia and northern Europe and winter sea-ice may have extended to the latitude of Britain in the eastern Atlantic.

Another impact of ice on the global environment is its effect on sea level. Eighteen thousand years ago Britain was joined to western Europe, Australia and New Guinea were one large island, and the Black Sea was isolated from the Mediterranean. The volume of water stored in continental ice sheets resulted in sea level being 120 m lower than today. If all the ice presently on the Earth's landmasses[6] melted sea level would rise by 80m, flooding most coastal regions. Sea level has been rising this century at the rate of about 1.5 mm/year, although about half of this rise is thought to be

---

[6] Sea-ice is floating on the ocean, as it is less dense than water, and only displaces its own weight of sea water. Therefore, if it melted the melt water would merely replace the volume of ice previously submerged.

Fig. 1.17. Average
boundaries of sea-ice at the
end of winter and summer
in (a) the Northern
Hemisphere, (b) (*opposite*)
the Southern Hemisphere.
The average winter extent
of icebergs is also shown,
and their source region in
(a). [Reprinted, with
permission, from J. G.
Harvey, *Atmosphere and
ocean: our fluid environment*
(London: Artemis Press,
1985), pp. 36–7, Fig. 4.6].

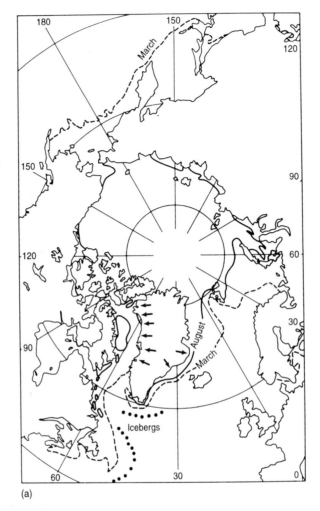

(a)

due to thermal expansion of sea water (as the global temperature has
increased slightly – see §1.8) rather than melting of land ice. In §§6.2.2 and
7.2.4 we will examine the phenomenon of sea level change in more detail.

The previous section demonstrated a significant climatic interaction of
polar ice, that of providing a mechanism for deep water formation. This
mostly occurs in the southern hemisphere (*austral*) winter under the *shelf ice*
of the Weddell and Ross Seas, off Antarctica. Here the ice can be tens of
metres thick.

Sea-ice thermally insulates the ocean from the atmosphere. It also
decouples the ocean from direct driving by the wind. *Pack ice* essentially
flows in the same direction as the underlying ocean; shelf ice, however, is
also subject to motion induced by its contact with continental shelves and
land glacier forcing.

The extent of oceanic ice cover varies dramatically with the season. Fig.
1.17 shows seasonal extremes for the two hemispheres. Comparing Fig. 1.15
with 1.17 shows the impact of the ocean circulation on ice distribution in

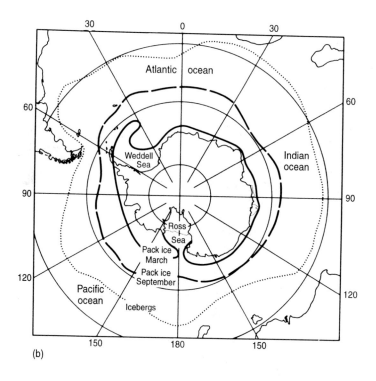

(b)

areas such as east of Greenland, where the local current pushes ice much further south than elsewhere.

1.5 The biosphere

The warming of the atmosphere provided by the greenhouse effect allows the biosphere to exist. Part of the biosphere consists of the biological activity in the ocean. This will be considered in detail in Chapter 4; the present section is concerned with the interaction of life on the terrestrial surface and the atmosphere.

The land biosphere provides many of the natural sources of greenhouse gases for the atmosphere, particularly carbon dioxide, methane and nitrous oxide. It can be thought, therefore, to contribute to its own existence[7]. Carbon dioxide is a product of respiration, as oxygen is a product of photosynthesis. Photosynthesis itself allows the terrestrial biosphere to act as a sink for CO_2 by storing it in each year's growth. This cycle will be discussed in detail in Chapters 3 and 4. Methane (CH_4) is another important greenhouse gas with significant natural sources. It is produced in *anaerobic* (oxygen poor) conditions by bacterial decay. This often occurs within bodies of standing, stagnant water, such as bogs, tundra, and rice paddies. It also occurs as a by-product within the digestive systems of animals (enteric fermentation), particularly ruminants such as cattle. Termites also produce CH_4 in their digestion of plant material. The

---

[7] This concept of the interdependency of the atmosphere and biosphere lies at the heart of the Gaia hypothesis, postulated by Lovelock (see also §1.2).

Fig. 1.18. Variation in
atmospheric transmission
of solar radiation to
Mauna Loa Observatory,
Hawaii for two years after
the eruption of Mt.
Pinatubo in June 1991. The
observatory is situated on
a mountain top, and is
therefore in the
mid-troposphere. The
decrease in solar
transmission shows the
reflection of energy by
particles trapped in the
stratosphere which take
two years to settle out.

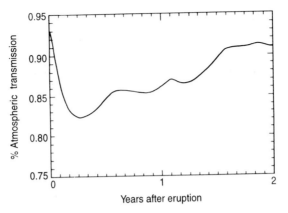

production of CH_4 due to water-supported anaerobic decay is two to three times as great as the contribution from enteric fermentation, which is about twice as important as termites. Nitrous oxide is a product of aerobic release of nitrogen within soils.

The land biosphere is also a component of the hydrological cycle: vegetation cover moderates the rate at which precipitation enters the soil, and determines the level of transpiration and photosynthesis. It also provides a way for momentum to be lost to the atmosphere by providing additional drag to the lower troposphere – the *planetary boundary layer* – as the wind blows over and through vegetation. This additional drag is a strong function of the type of vegetative cover.

1.6   The geosphere

The geosphere consists of all the solid Earth, from the soil surface, directly in contact with the atmosphere, to the mantle. It is the long-term reservoir of most of the compounds and elements which contribute to the climate system.

Direct connection between the interior of this store and the fluid envelopes of the atmosphere and ocean occurs in volcanic eruptions and hydrothermal activity. The former can, if the eruption is large, transport material and gases into the stratosphere. These can interact with the ozone layer, but more importantly, provide an effective screen to solar radiation, reducing the energy available to the climate system by reflecting back a substantial part of the incoming radiation. For example, the 1991 eruption of Mount Pinatubo in the Phillipines reduced the incident radiation at Mauna Loa, a mountain top observatory in Hawaii, by just over 10% for over a year (Fig. 1.18). Similarly, the 1982 eruption of El Chichon, in Mexico, caused a decrease of over 15% for a few months.

Historically, there have been numerous occasions when volcanic eruptions caused short-term cooling of the atmosphere. The eruption of Mt. Agung, in Indonesia in 1963 was probably linked to global cooling in the next year; Krakatoa in 1883 seemed to have had a similar effect. It is important, however, to note that not all large eruptions have such effects: the eruption of Mt. St. Helens in 1980 blew out a substantial proportion of

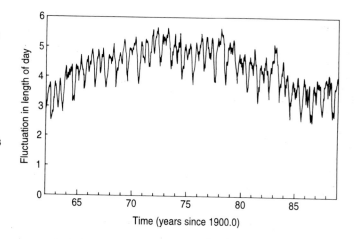

Fig. 1.19. Fluctuation in
the rotation period, or
'length of day' of the Earth
over a 30 year period. The
vertical scale is the
difference, in milliseconds
(10^{-3} s), from some
average rotation period.
The high frequency
oscillations are
atmospherically driven, for
instance the peak in 1982–3
is due to the decrease in
tropical easterlies allowing
the planet to rotate faster
during the 1982–3 El Niño
(see §5.2). The longer,
decadal scale, oscillation is
thought to be due to
motion within the Earth's
interior. [Data from Hide
and Dickey, 1991.]

the mountain, but the blast was directed sideways and most of the ash was put into the troposphere and quickly washed out in precipitation.

The geosphere defines the geographical boundaries of land, sea and air. We will see repeatedly how these substantially modify the radiative and circulatory climate of the atmosphere and ocean. One major impact that the geosphere has on climate is through geological change in these basic boundaries (continental drift). This will be considered in §1.8, and in more detail in Chapter 6 but the basic mechanism is contained in the theory of *plate tectonics*. This proposes that the upper region of the Earth, the lithosphere (the crust and the upper mantle), is composed of a number of basically stable platforms that move slowly, at the rate of up to a few centimetres per year, over the surface of the planet. These platforms, or plates, are believed to move in response to convection deeper within the Earth's interior where pressure and temperature are so high that the rock is ductile, or even fully liquid. Evidence for this interior convection is suggested by changes in the rotation period of the Earth of a few milliseconds over several decades. This would occur because of changes to the mass distribution, and hence the *angular momentum*, of the planet (see §2.5.2 and equation (2.9) for a discussion of angular momentum). It is known that changes in the atmospheric circulation can alter the rotation period by 1–2 milliseconds; the larger changes observed (Fig. 1.19) are attributed to such mass re-distribution within the Earth.

This plate movement is also associated with creation and destruction of plate material (Fig. 1.20). Molten material from the Earth's lower mantle is thought to upwell into the crust, or lithosphere, along mid-ocean ridges, forming new oceanic crust which then slowly spreads out at speeds of up to 60 mm per year. This may sound extremely slow, but it means that no oceanic floor is more than about 150 million years old. To conserve the surface area of the earth, destruction of crust must take place elsewhere. This also occurs in the ocean, in extremely deep submarine trenches. Here, oceanic crust is subducted beneath another, less dense, part of the lithosphere, typically a continental plate, and is forced to plunge steeply towards the mantle, being melted as it does so. The resulting narrow, but

Fig. 1.20. Schematic of a
subduction zone between
two oceanic plates. These
are being forced towards
each other by motion
within the deepest layer
shown. The leftmost plate
is less dense than that on
the right and so the crust
and upper mantle (top two
layers) of the right plate
are pushed under the left
plate. Melting of this
subducting material allows
the upwelling of magma to
support volcanoes in the
less dense plate.

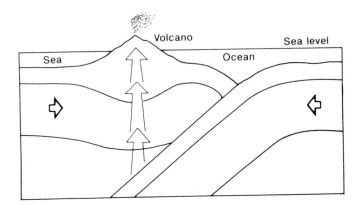

Fig. 1.21. Generalized
global distribution of
major earthquakes since
1800. Most earthquakes are
associated with modern
plate boundaries or are
re-activations of old plate
boundaries. [Fig. 5.2 of
Howell (1993), *Tectonics of
suspect terranes*.
Reproduced with
permission of Chapman
and Hall.]

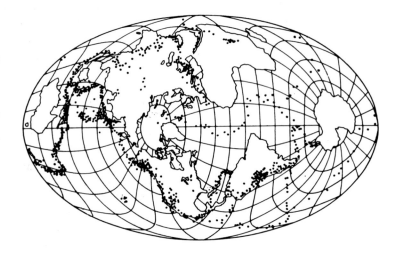

deep, trenches also form lengthy features on the ocean floor, practically
encircling the Pacific Ocean, for instance.

Subduction zones are volcanically active, because of the energy involved
in the subduction process (Fig. 1.21). Island arcs are associated with these
subduction zones, such as the Hawaian Islands in the Pacific, the Maldives
in the Indian Ocean or the Scotia Ridge region of the South Atlantic sector
of the Southern Ocean. Some of these arcs, such as the latter two examples,
are directly associated with subduction zones, being the stable crust's
response to being undercut. Others denote the former position of such
zones, either of existing zones which have moved to a new position, or ones
which have disappeared completely.

Plate movement and formation may also occur where two plates diverge
or where an existing plate breaks up due to internal stresses. The Red Sea,
with the Rift Valley in East Africa, is thought to be an ocean basin in the
process of formation. The Red Sea has already experienced sufficient
fissuring for two new plates to be identified, while the Rift Valley is in an
earlier state of fracture.

Fig. 1.22. Schematic of the different processes that affect the climate system, and their timescales.

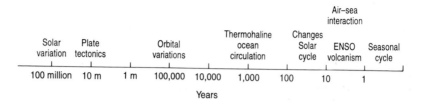

1.7   Timescales and feedbacks

The climate system is complex not just because within it a large number of processes and exchanges occur, but because different components act on widely differing timescales (Fig. 1.22) and interact in highly non-linear ways. The atmosphere, particularly the lower troposphere, has a large diurnal cycle. It reacts to changes in the basic heating source very rapidly. Within a few days transport and mixing can occur over continental scales, and within a few weeks, globally. In contrast, the biosphere's basic cycle is annual, as it responds to the change from winter to summer, or, in the tropics, wet season to dry. A diurnal cycle exists, but its significance is much less than in the atmosphere. It represents transitions from activity to dormancy rather than the annual cycle of growth and decay.

The other parts of the climate system act on much longer timescales. The ocean has a weak diurnal signal in its near-surface temperature, but this extends to only a few metres. Even the annual cycle in temperature only penetrates beyond a hundred metres in polar and mid-latitude regions. Large-scale mixing within the ocean occurs over months to years. Local eddies, the weather systems of the ocean, may exist for over a year; flow about the North Atlantic sub-tropical gyre takes two to three decades. The deep circulation imposes the absolute scale for over-turning and ventilation of the ocean. It may take several hundred, perhaps a thousand, years for water which sank in the Norwegian Sea to eventually re-surface in the Pacific Ocean.

The cryosphere changes on two timescales. There is advance and retreat of sea-ice margins in response to the annual cycle in solar radiation. Glaciers and ice sheets take much longer to alter substantially. Large-scale advances and retreats of the major ice caps occur over about a hundred thousand years, with smaller scale perturbations on scales of thousands of years (but see §6.2.2). Alpine glaciers tend to change on intermediate scales to these, altering their length by kilometres in tens of years.

The longest timescales occur within the geosphere. Diurnal and seasonal temperature signals only penetrate a few metres, at most, into the soil, although the response at the surface is large and strongly coupled to the incident radiation because of the low specific heat of the top-soil. Underground water movement occurs on these timescales to depths of hundreds of metres. The basic exchange of elements in the various cycles, however, occurs over hundreds of millions of years. Continents move at millimetres per year, creating new landforms in tens of millions of years.

These enormous differences in timescale are very important for the climate system. The atmosphere reacts quickly to change, but the other

Fig. 1.23. Schematic of
some of the many
processes involved in the
coupling of clouds to
climate.

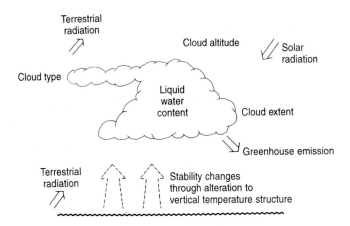

Fig. 1.23. Schematic of some of the many processes involved in the coupling of clouds to climate.

components act as buffers, retarding change. This buffering can delay unwanted alteration, but it also slows transitions to less extreme states.

There are also feedbacks within the system. We have seen how some components interact; for example, the biosphere and the atmosphere. However, these interactions, or changes within a component, may create conditions favourable, or unfavourable, to such processes continuing. This accentuation, or damping, of processes is known as positive, or negative, feedback. An illustration of positive feedback you have probably experienced is the magnification of noise that can occur in a broadcasting system when too much sound from loudspeakers enters the microphone. This can create a build-up of sound quickly as the signal continually feeds from the speakers to the microphone in ever-increasing volume, until the microphone is redirected to break the sound conveyor belt.

An example in the atmosphere of such a positive feedback is contained within the hydrological cycle. As water is evaporated the water vapour adds to the greenhouse effect, warming the atmosphere. This warming promotes further evaporation from the land or sea, and the cycle continues, producing greater and greater warmth. Eventually, as in most positive feedback processes, another process reacts to this amplification by dampening, and finally reversing, the initial mechanism. In the cycle just described negative feedback mechanisms are linked to the growth of clouds. Clouds will form when enough water vapour has accumulated in the atmosphere. These are powerful reflectors of solar radiation, thus preventing much primary solar fuel from entering the positive feedback process. The water vapour/cloud interactions are even more extensive than the simple example just discussed. They are extremely complex, and not fully understood. Fig. 1.23 conveys an impression of this complexity.

Another area in which feedbacks are strong is in the links with the biosphere. It is only just becoming clear how extensive these are, in both a positive and negative sense. Chapter 4 will examine some of these newly discovered interactions.

1.8 Variation of the climate system over time

The early atmosphere of the Earth was very different from the present mixture of mainly nitrogen and oxygen. Considerable quantities of hydrogen and helium were present, which have mostly escaped into space.

Much of the present atmosphere originates from volcanic emission of gases and subsequent photo-dissociation, or reaction with solar radiation. The existence of life on Earth has also contributed towards the present atmospheric composition by providing much of the oxygen. Stronger solar radiation, and a faster rotation rate, have also influenced the atmosphere during its formation.

Since the advent of the present mixture of atmospheric gases the main determinant of the basic global climate pattern has been the Earth's geography. The strength of the solar radiation has probably been essentially constant since the present atmospheric composition was attained, except for the comparatively small variations caused by the periodicities in the Earth's orbit, discussed below. Much of the atmosphere's driving energy is provided by the underlying surface, whether land or sea. Thus, if the present continental layout was changed, either in position or the relative proportions of land to ocean, then this heating would also change. In §1.6 we saw how the plates that make up the Earth's crust collide to form mountains and oceanic troughs, or divide to form new oceans. This occurs sufficiently slowly for little variation to be observed for several million years but such movement leads to very long-term climatic evolution.

Over the past 300 million years the Earth has experienced three prolonged periods of extensive glaciation. Two of these are 250 and 600 million years ago, at pronounced minima in the global temperature record shown in Fig. 1.24. We are currently in the third of these glacial epochs, although in an *interglacial* intermission in a longer period of predominantly glacial conditions. For 210 million years after the last glacial era the climate of the Earth was much warmer than today. This climatic evolution probably occurred because of tectonic plate movement. A major difference between the glacial and warm eras is whether there is land over one or both of the polar zones. If there are polar land masses there is then the opportunity for snow to accummulate during such region's sun-less winters, and the depth of polar coldness for this snow to last through the summers and eventually form ice sheets.

During the Cretaceous period, while dinosaurs still dominated the Earth, the continental plates began to 'drift' towards their present positions, from a previous, more consolidated, state. The continental geography at the beginning of the process, 100 million years ago, is shown in Fig. 1.25. A zonally averaged temperature distribution for this state is given in Fig. 1.26, with the modern distribution, and a full glacial state from 21 000 years ago.

As time progressed, most of the continents moved north, although at different rates. Australia and Antarctica divided rather later, about 30 million years ago, and Antarctica drifted southwards while Australia travelled northwards. As Antarctica drifted over the South Pole, and Australia moved sufficiently far north for a circumpolar current to develop

Fig. 1.24. Mean global
climate variation over the
Earth's history. The left
column is mean global
temperature and the right
column is mean global
precipitation. Note the
non-linear timescale. [Fig.
9.1 of Frakes (1979),
*Climates through geologic
time.* Reproduced with
permission of Elsevier
Science Publishers.]

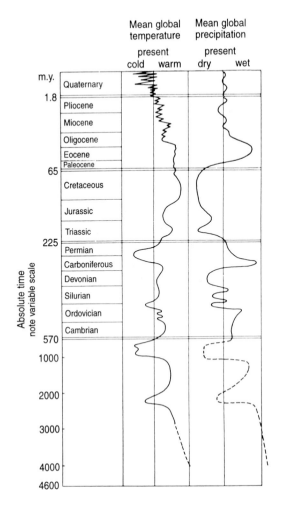

and effectively isolate Antarctica from poleward oceanic heat transport, ice sheets began to form. This occurred 30–40 million years ago. The slow global cooling, quickened by the formation of the Antarctic ice cap, was coupled to the slow isolation of the Arctic Ocean by North America and Eurasia and the raising of the Tibetan plateau by the collision of India with Asia. Eventually this Arctic isolation, in conjunction with winter cooling of extensive land areas poleward of the Arctic Circle and deflection of the global atmospheric circulation by the Himalayas, led to the development of Northern Hemisphere ice sheets.

During the last million years the Earth has experienced repeated extensive glaciation in its Northern Hemisphere. The continental geography has been essentially constant during this time. Only minor variations in the distribution of land and sea, due to sea level oscillations of up to 150 m, have occurred (this will be discussed further in §6.2.2). A reason for this glacial cycling needs to be sought elsewhere.

The earth's orbit about the Sun is not circular. Instead it is slightly elliptical, with the Earth being at one of the focii of the ellipse. The closest

Fig. 1.25. Palaeogeographic reconstruction at 100 million years ago, after the super-continent, Pangaea, has begun to break up. Light areas on continents indicate regions flooded by shallow seas (maximum depth 100–200m). The narrower part of the region of continuous ocean stretching around the globe in tropical regions is sometimes called the Tethys Seaway. [Fig. 3 of Barron *et al.* (1980). Reprinted with permission of Elsevier Science Publishers.]

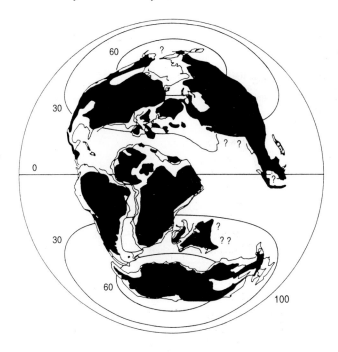

Fig. 1.26. Zonal mean temperature during a warm Earth climate (100 million years BP, data from Barron and Washington, 1984) – dashed line; a glacial climate (21 000 years BP, data courtesy of Paul Valdes) – dotted line; and the present interglacial climate.

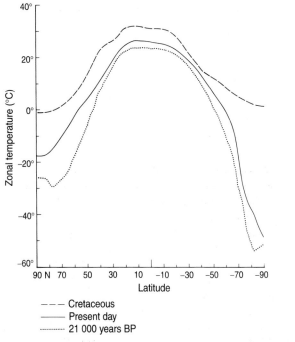

point to the Sun, or *perihelion*, is reached in mid-December and the furthest point in mid-June (see Fig. 1.27). In addition, the Earth is tilted at an angle of 23.5° to a line perpendicular to the plane of rotation about the Sun. This orbit is not permanent. There are slight periodicities in its components which alter the distance, or orientation, of the Earth from the Sun. The

Fig. 1.27. Periodicities of
the Milankovitch cycle in
the Earth's orbit. The
Earth's position
corresponds to Northern
Hemisphere summer.
[From Bigg, 1992c]

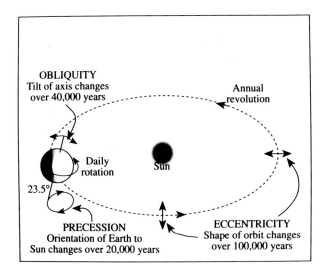

eccentricity of the ellipse varies from 0.0 (essentially circular) to 0.06, with a
period of 100 000 years, although there is a modulation to this cycle so that
the spread of eccentricity for any particular cycle is usually well within this
extreme range. The tilt of the Earth's axis varies by 2.5° over 40 000 years.
Finally, the position of the perihelion moves through the year (the orbit is
said to precess) with a period of 20 000 years. The variation in eccentricity
alters the total amount of solar radiation to reach the Earth's orbit over a
year, the change in tilt changes the latitudinal variation, and the precession
of the perihelion alters the seasonal distribution of radiation. All of these
periods appear in sedimentary records of the past million years; the
eccentricity shows up most strikingly in the recent global record as the
100 000 year oscillation in Fig. 1.23. Prior to about 470 000 years ago glacial
cycling occurred with a 40 000 year periodicity, still remarkably similar to
one of the orbital periods. In Chapter 6 we will consider what might have
caused this change of period.

The glacial fluctuations mainly affect the northern hemisphere. Ice sheets
have extended as far south as southern England in Europe and the central
United States in North America. There is a lag between the astronomical
parameters and the climate due to the strong feedback between the highly
reflective ice sheets and low radiation, once the ice sheets form. There is also
a link between the state of the glaciation and the concentration of key
greenhouse gases, such as CO_2 and CH_4. Chapter 6 will discuss the strong
impact these events had on the ocean circulation, and the role of the ocean
in assisting, and prolonging, such events.

The peak of the last glaciation was reached about 18 000 years before
present (BP) but the deglaciation has not been a monotonic process since
then. After a slow warming initially, the globe rapidly shed its northern ice,
predominantly in two periods of melting about 12 000 and 9000 years BP.
Global climate 9–6 000 years BP was as warm as, and sometimes warmer
than, today (Fig. 1.28). Even over the past 2000 years the global temperature has varied by up to 1 °C on either side of the present value. The Vikings

Fig. 1.28. Variation of Northern Hemisphere mid-latitude temperatures since the peak of the last glaciation. Note the pronounced cooling around 11 000 years BP (the Younger Dryas) and that the warmest climate was some 6–7000 years BP.

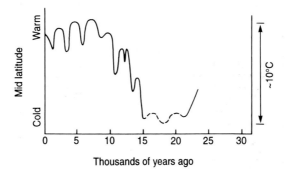

Fig. 1.29. The variation of the atmospheric concentration of CO_2 since 1860 (solid line), plotted with the mean global temperature over the same period. The bars show the year to year variation in global mean temperature, relative to the 1951–80 average. Note the significant rise in global temperature from 1920–40 and again from 1975–90; in contrast the rise in atmospheric CO_2 occurs continuously.

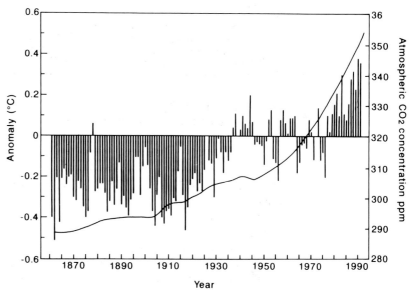

were able to settle Greenland, Iceland and part of North America in a milder climate than at present. In contrast, during much of the period from 1300 to 1800 significantly colder weather than today was experienced in western Europe. The winters, in particular, were extreme, with frequent freezing of major rivers such as the Thames. These changes are poorly understood. Extreme volcanic activity could be responsible for periods of cooling, but the linking evidence is not clear. The ocean could also have a role. Chapter 6 will explore these possibilities further.

Today the principal cause of any imminent climatic change is thought to be increases in atmospheric concentrations of greenhouse gases produced by man since the beginning of the Industrial Revolution in the eighteenth century. Such increases should magnify the greenhouse effect, warming the Earth.

In Fig. 1.29 the evolution of global temperature over the last century is shown. This figure also shows the change in atmospheric CO_2 concentration over the same interval. The latter shows an almost smoothly increasing curve. The global warming of about 0.5°C since 1880 has, however, mainly occurred during two short periods – 1920 to 1940 and 1980 to 1990. At other times the global temperature has been roughly static. Therefore any link between greenhouse gas concentration and global warming is far from linear. As we have briefly seen in this chapter, the climate system is complex and highly interactive. The amount, and rate, of any warming will therefore be dependent on the inter-relationship of the different components of the system – natural variability and feedbacks induced by greenhouse warming – and not just changes in the composition of the atmosphere itself. Once we have explored the mechanisms of interaction, focussing on those involving the ocean, we will return to the climatic future in Chapter 7.

Further reading

A complete reference list is available at the end of the book but the following is a selection of the best books or articles to follow up particular topics within this chapter. Full details of each reference are to be found in the Bibliography.

Brown *et al.* (1989): An excellent introduction to basic physical oceanography.

Chahine (1992): An excellent discussion of the hydrological cycle and its climatic importance.

Crowley and North (1991): An extremely readable, and well referenced, discussion of climates of past times.

Hartmann (1994): Excellent discussion of basic meteorological and radiational elements of climate and physical feedbacks.

Harvey (1985): A clear and simple discussion of basic oceanography and meteorology.

Houghton *et al.* (1990): Careful discussion of the evidence for current and future climatic change.

McIlveen (1992): Good discussion of basic meteorology.

2 *Physical interaction between the ocean and atmosphere*

The ocean and the atmosphere share a common boundary: the air–sea interface. This direct physical contact enables the two fluids[1] to exchange energy and matter. In this chapter we will examine these exchanges from a physical perspective, leaving a discussion of the chemical controls on matter, particularly gas, transfer to Chapter 3.

Physical interaction between air and sea takes a number of forms. We are all familiar with the production of waves on water, due to wind blowing across the surface. We have also all seen that strong winds tend to force these waves to break, creating whitecaps. When these whitecaps disintegrate, salt water is sprayed into the atmosphere. The addition of fresh water to the ocean through precipitation is another obvious link between the two media.

There are also a number of less obvious interactions. The ocean is given energy by solar radiation passing through the atmosphere. Infra-red radiation from within the atmosphere enters the ocean; it also leaves the ocean to warm the atmosphere. There are also other mechanisms for heat exchange. The direct physical contact of the sea and air means that there is exchange of energy through collision of molecules in the surface layer of each fluid. This energy is known as *sensible heat*. There will also be an exchange of the molecules themselves, generally resulting in net evaporation, and therefore transfer of latent heat, from the water surface.

The momentum transfer that occurs when the wind blows over the ocean results in motion within the water. Locally, this causes small-scale flows to develop. On larger scales, the net effect of the wind over entire ocean basins is to drive the general circulation pattern described in §1.3.2.

The converse, a driving of the atmosphere by the ocean, also occurs. This is not so much due to the physical drag, as the exchange of heat energy. Local changes in atmospheric density and water content occur because of contact with the ocean beneath, producing larger scale pressure gradients, and providing latent heat to assist the development of active weather systems. As these heat exchanges are the key to understanding the impact of the ocean on climate we shall begin by examining these processes in detail.

---

[1] The atmosphere is often called a fluid because it flows and obeys similar physical laws to conventional fluids, such as water.

Table 2.1. *Reflectance of radiation from a calm water surface (after Jerlov, 1976)*

Incident angle (° from vertical)	Reflectance (%)
0	2.0
10	2.0
20	2.1
30	2.1
40	2.4
50	3.4
60	5.9
70	13.3
80	34.9
90	100.

2.1   Radiation

2.1.1   *Solar radiation*

In Fig. 1.11 we saw that 51% of incident solar radiation reaches the Earth's surface; on average, about 12% of this is reflected back into space, the proportion varying according to the albedo of the surface. Over the oceans most of the incident solar radiation enters the water, although the albedo depends strongly on the angle at which the Sun's radiation hits the surface. Table 2.1 gives the *reflectance* of direct sunlight, illuminating a calm water surface, over a range of angles of incidence.[2] This should be contrasted with the albedo of some typical terrestrial surfaces, given in Table 7.2. Until acute angles are reached well over 90% of the radiation passes through the surface. Thus, over most of the globe, for most of the daylight hours, almost all direct solar radiation incident on the ocean surface enters the water.

There are several complications to this apparently simple picture. Solar radiation suffers significant *scattering* by air molecules on its passage through the atmosphere. This is known as *Rayleigh scattering* and occurs because the molecules are small in comparison to the wavelength of the radiation (10^{-10} m compared to 5×10^{-7} m). The degree of scatter is inversely proportional to the fourth power of the wavelength so lower, or blue, wavelengths are preferentially scattered. This scattered radiation is spread through the atmosphere, creating our perception of the sky being blue, and reaches the surface in a range of incident angles. Because of this range rather more scattered, compared to direct, light is reflected from the sea surface: 5–6%, compared to 2–3%, in most circumstances.

The presence of clouds does not significantly alter this proportion, although it does mean that a greater percentage of the radiation that reaches the surface is scattered radiation. On cloudy days, therefore, perhaps twice as much radiation will be reflected from the sea surface as on clear days.

[2] Reflectance and albedo are not identical because some of the radiation entering the water can be forced, through scattering, to re-enter the atmosphere. Reflectance measures the immediately reflected radiation from the surface, while the albedo is the proportion of the incident radiation that is lost from all internal and external reflective processes.

Fig. 2.1. Schematic of the
refraction of light passing
from air to water.

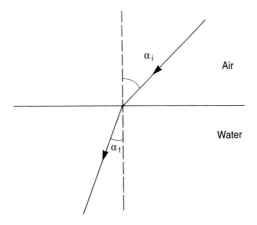

A further complication arises when the sea surface is not calm. The creation of waves means that the sea surface can present the full range of angles to the incident radiation. If the sun is high in the sky this tends to increase the mean angle of incidence, but the small dependence of reflectance on angle under such conditions means that the net effect on reflectance is negligible. However, at sunrise and dusk, and in high latitudes, rough seas tend to decrease the average angle of incidence. This can decrease reflectance by a factor of two and so allow significantly more radiant energy to enter the ocean.

A final consideration in determining the actual albedo of the sea surface is the fate of light that penetrates into the sea. A proportion of this will be scattered by the sea water back to the surface and will thus be able to pass into the atmosphere, supplementing the initial reflection. Countering this scattering loss, however, is the *refractive index, n*, of water. This gives the ratio of the speed of electromagnetic radiation in a vacuum to that in water. The ratio of n for two media through which light travels gauges the resistance to the passage of light from one medium to another, or the reflecting effect of the interface between the media. The refractive index for water is $n_w = 1.33$ and that of air is $n_a = 1.00$, so much more light is reflected when passing from water to air (n_w/n_a) than the converse (n_a/n_w). In addition, Snell's law for refraction:

$$\frac{\sin \alpha_i}{\sin \alpha_t} = \frac{n_w}{n_a} \tag{2.1}$$

where angles α_i and α_t are as shown in Fig. 2.1, has the consequence that, when light is reflected back towards the surface, there will be total internal reflection beneath the water surface when α_i (from below) is greater than 48.5°. Therefore, only radiation incident on the sea surface from within the ocean at angles less than 48.5° will contribute to the back-scattered component of the albedo; 48% of the under-water incident radiation will be reflected back into the water.

This means, however, that about half of the radiation that enters the sea is potentially able to escape back into the atmosphere, if all of this radiation

Fig. 2.2. Energy spectrum, and its variation with depth, in a beam of solar radiation penetrating the ocean in (a) the western sub-tropical Atlantic, (b) the Baltic. The solid lines show the spectrum at the labelled depth. Note the rapid attenuation with depth in the sediment-laden waters of the Baltic. Data from Jerlov (1976).

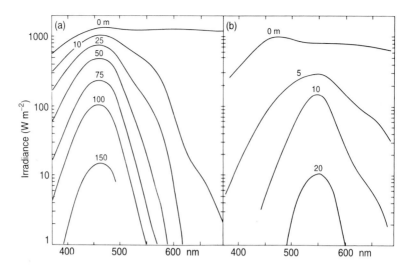

were scattered within the ocean back to the air–sea interface. In practice, almost all of this radiation is absorbed within the sea, converting electromagnetic energy into heat. Most of this absorption is carried out by water molecules. Dissolved salts absorb weakly in the ultra-violet, while suspended sediment and plankton absorb variable amounts of solar radiation, depending on their type and concentration. We saw in §1.2.1 that water vapour is a powerful absorber of the infra-red part of the spectrum. Fig. 2.2 and Fig. 1.9 show that significant absorption extends into the higher wavelengths of the visible portion of the spectrum. This means that a few metres below the surface of the sea the blue–green region of the spectrum has become more dominant because of selective removal of the higher, red, wavelengths. This is shown in Fig. 2.2. The sea therefore appears blue–green to an observer, as this is the light back-scattered out of the water (Fig. 2.3). The green, or murky grey, of coastal seas, such as the North Sea, is caused by high plankton population (except in winter, see Chapter 4), dissolved organic material and suspended sediment. Plankton *photosynthesise*, selectively absorbing wavelengths other than near the green segment of the visible spectrum to catalyse this reaction (see §4.1.1). Note that this is also why leaves are green! The high sediment concentrations in coastal waters tend to make absorption more uniform over the spectrum.

Fig. 2.2 also shows that less than 50 m below the surface the light intensity, even in the blue–green wavelength band, is reduced to a quarter, or less, of that just below the surface. In seas rich in sediment and plankton this decline in irradiance is even more pronounced (Fig. 2.2(b)). The *euphotic zone* is defined to be the depth at which light intensity is only 1% of the surface value. In the clear, biologically inactive, waters of the subtropics this depth can approach 100 m (Fig. 2.2(a)). In turbulent, muddy, and biologically rich coastal waters such as those of the North Sea, or Baltic, the euphotic zone may be less than 20 m deep (Fig. 2.2(b)). In Chapter 4 we will see how the concept of the euphotic zone dominates the biological productivity of the oceans.

Fig. 2.3. Energy spectrum
of solar radiation at the
ocean surface, but of that
scattered within the ocean
and re-entering the
atmosphere. The dashed
line shows a western
sub-tropical North Atlantic
spectrum, the solid line is
from the Baltic Sea. Note
the much reduced
scattering in the
sediment-laden waters of
the Baltic. Data from
Jerlov (1976).

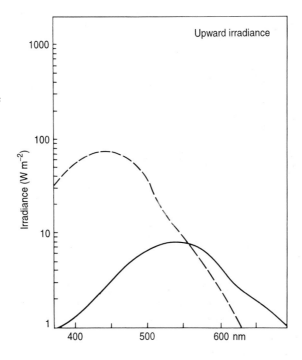

Fig. 2.3. Energy spectrum of solar radiation at the ocean surface, but of that scattered within the ocean and re-entering the atmosphere. The dashed line shows a western sub-tropical North Atlantic spectrum, the solid line is from the Baltic Sea. Note the much reduced scattering in the sediment-laden waters of the Baltic. Data from Jerlov (1976).

2.1.2 *Long-wave radiation*

The sea receives solar radiation from the Sun. It also receives long-wave, infra-red, radiation from the atmosphere. In addition, it emits long-wave radiation to the atmosphere. The balance between the energy gained by solar radiation and incoming infra-red, and the energy lost through outgoing infra-red determines the radiation budget of the sea surface.

In §1.2.1 we saw how the greenhouse effect absorbs infra-red radiation headed towards space, and re-radiates much of this energy back towards the Earth's surface. The amount of long-wave radiation lost by the ocean will therefore be strongly dependent on the water vapour concentration, this being the principal, and fastest varying, greenhouse gas. Clouds, as liquid water, will also absorb some of the infra-red radiation emitted by the ocean. The determining factor in the amount of long-wave radiation emitted by the sea into the atmosphere is, however, its surface temperature, T_S. To the atmosphere the ocean appears as a body at temperature T_S radiating according to the Stefan–Boltzman Law (1.1); the thermal structure beneath the surface is only visible in its surface signature. The infra-red radiation from the sea surface, the long-wave heat flux Q_B (measured in Wm^{-2}), is therefore given by (1.1), but reduced by the returning greenhouse emission from the atmosphere. This function can be empirically estimated:

$$Q_B = 0.985\sigma T_S^4(0.39 - 0.05\sqrt{e_a})(1 - 0.6n_c^2) \qquad (2.2)$$

where e_a is the *vapour pressure* (measured in millibars) – the pressure exerted by the weight of water vapour molecules present in a column of air, one

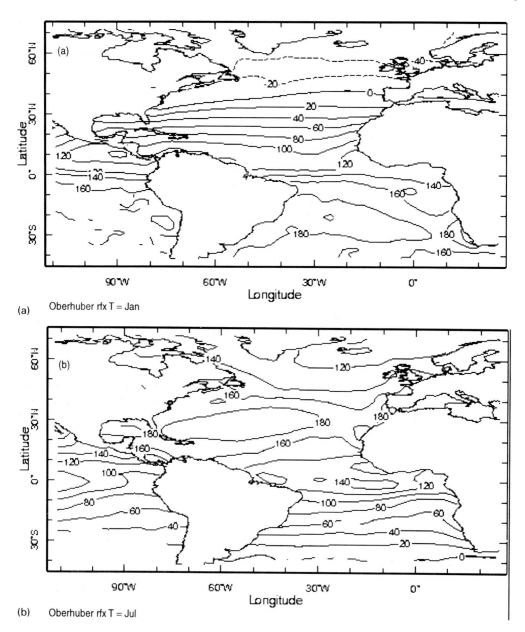

(a)   Oberhuber rfx T = Jan

(b)   Oberhuber rfx T = Jul

Fig. 2.4. Net radiation (in Wm^{-2}) at the ocean surface over the Atlantic in (a) January, and (b) July. The contour interval is 20 Wm^{-2}. Dotted contours indicate negative, or net outgoing, radiation. Data from Oberhuber (1988).

metre square, above the surface – of the atmosphere (taken above the standard observing height of 10 m) and n_c is the proportion of cloud cover. The vapour pressure can vary by a factor of ten, depending on the atmospheric conditions. Typical values are of the order of a few tens of millibars; a typical atmospheric pressure is a little over 1000 mb. The correction factors for the trapped, or return, long-wave emission from clouds and greenhouse gases reveal the importance of water vapour for this process. A completely cloud covered sky reduces the loss of radiation by 60%. This is almost the same as is absorbed by all greenhouse gases, save water vapour, in the absence of clouds: 61%. Note also that in polar regions, where there is little water vapour in the air, the effect of water vapour is only an additional 10–15% on top of the other greenhouse gases, while in the tropics it can be as much as 25%[3], or almost a third of the total non-cloud return long-wave radiation.

The average energy of the net short-wave and infra-red radiation received by the Atlantic Ocean is shown for January and July in Fig. 2.4, calculated from such empirical formulae as (2.2) and a similar one for the incident short-wave radiation. In the winter hemisphere there is net loss of radiation poleward of 35–40° and the distribution is essentially zonal. In the summer hemisphere, and near the equator, variation in cloudiness has a strong impact, for instance it leads to the minima in net radiation over the central North Atlantic off western Europe (July) and the eastern equatorial Atlantic (January and July).

2.2 Heat exchange through latent and sensible heat

2.2.1 *Latent heat*

Radiation dominates the exchange of heat between the atmosphere and ocean. Other physical mechanisms do, however, also contribute to the net heat flux. The most important of these is latent heat transfer. When water is evaporated from the ocean surface energy is supplied to the molecules to free them from the strong inter-molecular bonds within liquid water. This process was discussed in §1.3.1. When the water molecules condense to form water droplets, usually in clouds, this energy is released to heat the surrounding air. This internal source of heat adds to a cloud's buoyancy, allowing it to penetrate higher into the troposphere. Such additions of heat can eventually assist in the creation of new pressure gradients within the atmosphere, thus driving large-scale atmospheric motion.

Latent heat transfer is therefore an important means of re-cycling energy through the ocean–atmosphere system. A peculiar property of the atmosphere – possible *saturation* with respect to water vapour – makes this form of heat exchange highly temperature-dependent. A parcel of air said to be saturated with water vapour is one where its constituent water vapour is in equilibrium with an underlying flat surface of water. This means that the same number of water molecules enter the body of water from the air, per second, as are evaporated from the surface. We have already implicitly assumed knowledge of this property in discussing condensation; air containing more water vapour than its saturated level tends to form clouds, with the water vapour condensing to form droplets (the climatically important process of cloud formation, including the influence of the ocean, will be discussed in more detail in §2.8.3, §4.4, Chapters 5 and 7).

The variation with temperature of the saturation level is shown in Fig. 2.5, in terms of the saturated *specific humidity*. At temperatures near 0°C only a few grammes of water vapour, per kilogramme of air, need to be present in the atmosphere before saturation is reached, while at 30°C 30

---

[3] Note from (2.2) that the above proportions are multiplicative, not additive.

Fig. 2.5. Variation of saturated specific humidity with temperature at a pressure of 1000mb. [From Bigg (1992b)].

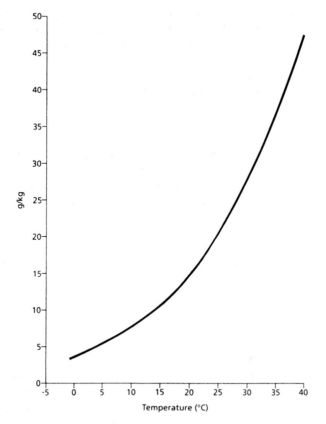

grammes are required. Air in the tropics, therefore, can contain much more vapour than air in polar latitudes. Tropical weather systems can thus be much more energetic, because of the greater potential source of latent heat. In §2.11.1 we will see that hurricanes are the most dramatic consequences of this tropical latent heat excess, but Chapter 5 will show that this tropical energy release can significantly influence the climate of the *extratropics* as well.

The latent heat, Q_E (Wm^{-2}), added to the atmosphere by the ocean is

$$Q_E = L_v E \tag{2.3}$$

where E is the evaporation (in kgm^{-2}s^{-1}) and L_v is the latent heat of vapourisation. The latter is the energy required to vapourise one kilogramme of water. It depends slightly on temperature, but for our purposes can be taken as $L_v = 2.5 \times 10^6$ Jkg^{-1}.

The determination of E is more difficult. In the first few millimetres above the ocean surface the air is in equilibrium with the underlying water surface: it is saturated with respect to water vapour. The evaporation rate will therefore depend on the humidity gradient between this saturated micro-layer and the planetary boundary layer as a whole. Conventionally, the specific humidity at 10 m is assumed to represent that of the lower boundary layer atmosphere. The evaporation rate will also depend on both the degree of turbulence and strength, u, of the wind, as the wind transports fresh supplies of air to the location being studied. Once again it is extremely difficult to measure evaporation at sea (it is more complex than measuring the depth of water lost from a pan over land as well!), so an empirical formula must be used for its estimation:

$$E = c_E \rho_a u (q_s - q_a) \tag{2.4}$$

In (2.4) c_E is a non-dimensional number approximately equal to 1.5×10^{-3}, ρ_a is the density of air, q_s is the saturated specific humidity at the sea surface (where the air is at the sea surface temperature, T_s), and q_a is the specific humidity at 10 m above the sea surface (q_s and q_a are measured in kg/kg). This latter variable is unlikely to be at saturation for the air temperature, T_A.

Latent heat transfer is extremely variable, both in space and time. It can be near zero in still, foggy conditions or comparable with the radiation terms in dry, warm, and windy weather. Average evaporation rates for the Atlantic Ocean in January and July are shown in Fig. 2.6. These reveal very different patterns to the net radiation of Fig. 2.4. Peaks in evaporation occur throughout the year in the sub-tropics because of the warm dry air in these zones of atmospheric subsidence (see §1.2). The strong evaporation off South Africa all year, off the eastern United States in January, and in the North-east Atlantic throughout the year occurs because of warm upper ocean currents. These are the Agulhas Current, the Gulf Stream, and the North Atlantic Drift respectively (see Fig. 1.15). The surface water in these currents is at generally higher temperatures than the overlying air, particularly in winter, so that the humidity gradient in the lower atmospheric boundary layer is enhanced. The frequent occurrence of strong winds adds to this intensification of evaporation, as does the supply of dry air off the neighbouring land masses. In Chapter 5 we will see that such strong forcing of the atmospheric heat budget by the ocean is important in both regional and global climate.

2.2.2   *Sensible heat*

The direct physical contact of the atmosphere and ocean enables energy to be exchanged between them by *conduction*. Such energy exchange is known as sensible heat. This occurs due to collisions between the molecules of the two fluids at their interface, with energy being transferred to the cooler, and therefore, slower, molecules. It should be noted that this process is a statistical one: the molecular speeds cover a range of values with their mean being the important quantity for determining temperature and energy transfer through conduction. Sensible heat transfer therefore depends on the temperature difference between the near-surface air and the sea surface. As with the latent heat flux calculation, the air's temperature at 10m is assumed representative of the near-surface region. Turbulence, and high wind speeds, encourage conduction by mixing air from higher in the

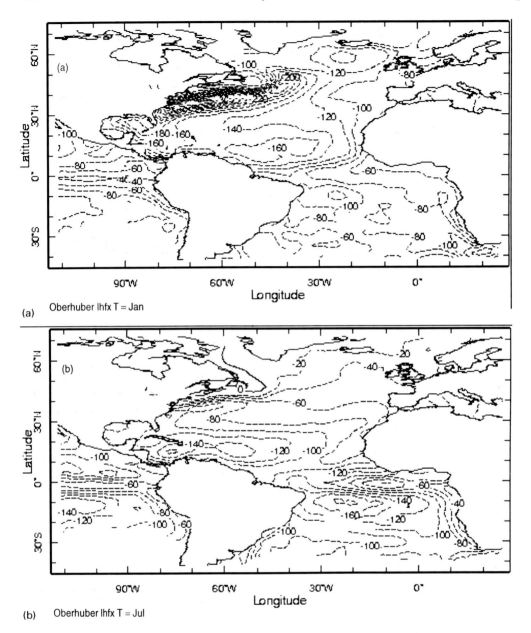

(a)   Oberhuber lhfx T = Jan

(b)   Oberhuber lhfx T = Jul

Fig. 2.6. Latent heat flux
(in Wm^{-2}) at the ocean
surface over the Atlantic in
(a) January, and (b) July.
Dotted contours, every 20
Wm^{-2}, indicate net loss of
heat from the surface. Data
from Oberhuber (1988).

atmosphere with that at the surface, allowing the ocean to interact with
more air than that in the shallow surface layer.

The sensible heat transferred to the atmosphere, Q_S (measured in Wm^{-2}),
is again difficult to measure so the empirical formula

$$Q_S = \rho_a c_H u (T_S - T_A) \qquad (2.5)$$

is often used. The Dalton number, c_H, is usually taken to be a function of the
degree of turbulence in the atmosphere. A typical range of values is from

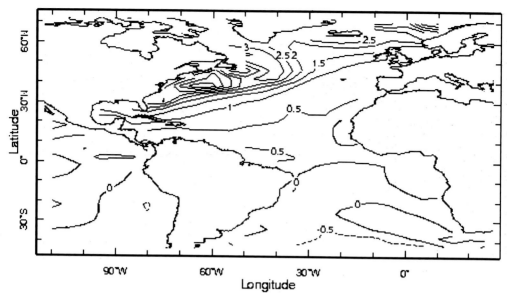

OSUSFC.DATA Z=0.0 T=Feb D=dtsa

Fig. 2.7. Mean February air–sea temperature difference (T_s - T_a) over the Atlantic Ocean. Contour interval is 0.5°C; solid, or positive, contours indicate a warmer ocean than atmosphere while dashed, or negative, contours indicate a warmer atmosphere. Data from Oberhuber (1988).

1.10×10^{-3} in an atmosphere with much vertical mixing to 0.83×10^{-3} in stratified air. The sensible heat transfer is generally much smaller than the other components of the heat balance at the air–sea interface because the temperature difference between the ocean and atmosphere is often less than 2°C. This is shown for the Atlantic Ocean in Fig. 2.7. Almost everywhere equatorward of 30° there is a sensible heat flux to the atmosphere smaller than 10 Wm^{-2}, in some places even negative, or towards the ocean. Higher values occur in some similar places to latent heat flux peaks, such as over warm ocean currents. This is particularly true in winter when the air–sea temperature difference is greatest and the winds tend to be stronger. Thus, these two processes reinforce in such areas to contribute substantially to the atmospheric heat supply.

2.3   The oceanic heat balance

The exchange of heat between the ocean and atmosphere principally consists of the four terms discussed above. Solar radiation provides the input to the ocean, and, in most situations, the net long-wave radiation, latent heat and sensible heat result in transfer of energy to the atmosphere. The net heat exchange, Q (in Wm^{-2}), at a location is therefore

$$Q = Q_I - Q_B - Q_E - Q_S \tag{2.6}$$

where Q_I is the incident solar radiation that is absorbed by the ocean and positive Q indicates addition of energy to the ocean. This expression neglects latent heat from the condensation of water onto the ocean surface and heat transfer due to mixing of precipitation with sea water. These contributions are both extremely small, however, and can be neglected in

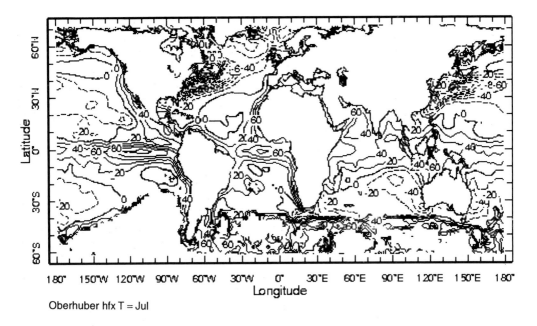

Oberhuber hfx T = Jul

Fig. 2.8. Annual average oceanic heat gain (in Wm^{-2}) over the global ocean. Contour interval is 20 Wm^{-2}; dotted (or negative) contours indicate a net loss of heat by the ocean. Data from Oberhuber (1988).

most circumstances. Note that if a full heat budget, rather than just the heat exchange, for a particular location in either the atmosphere or ocean was computed then the additional heat supplied (or lost) due to advection would need to be considered. This can be considerable, particularly for some of the regions of the Atlantic already highlighted.

An estimate of the distribution of the annual average air–sea energy exchange is shown in Fig. 2.8. There are extensive regions in which the ocean loses heat to the atmosphere. Referring to the surface ocean circulation shown in Fig. 1.15 it can be seen that these are principally regions with strong, poleward-flowing, warm currents, where we have already seen that latent and sensible heat fluxes increase. Typical areas are in the Gulf Stream of the North Atlantic, and the Kuroshio Current of the North Pacific.

If we examine the surface atmospheric circulation, shown in Fig. 1.6, it can be seen that these areas of heat input to the atmosphere tend to be close to consistent cyclone centres. The latent and sensible heat energy is responsible for recurring frontal development. We shall explore these associations further in §§2.11 and 5.1.4.

There are also areas in Fig. 2.8 where the ocean takes a considerable quantity of energy from the atmosphere. These tend to be close to the equator, and on the eastern margins of ocean basins. Upwelling of cold water from perhaps several hundred metres below the sea surface occurs in these regions. This reverses the sensible heat flux, lessens the long-wave radiation, and, because both the over-lying air and ocean surface are cooled, lessens the latent heat loss. This interaction will be discussed further in §§2.10.2 and 2.10.3, and Chapter 5.

The estimates in Fig. 2.8 are climatological averages. They provide a qualitatively correct picture of the global balance of heat exchange between ocean and atmosphere. However, there are considerable uncertainties within the basic observations, and also within the terms of the empirical formulae. In order to predict climatic change the terms in the heat balance need to be known to high accuracy, because small changes of a few Wm^{-2}, if consistent over a large region, can push the delicate balances within the climate system towards a new state. The net change to the long-wave radiation balance of the atmosphere due to a doubling of the CO_2 concentration is estimated to be only about $4\ Wm^{-2}$. The ramifications of this potential change, as will be discussed in Chapter 7, may be substantial, so accuracy in our estimates of the air–sea heat exchange is important.

There are a number of components of the empirical formulae that require more investigation. The transfer coefficients, c_H and c_E, are actually functions of wind speed or atmospheric stability, rather than constants. At low wind speeds the transfer processes are under-estimated because gusts of stronger winds produce short, but vigorous, periods of enhanced evaporation and conduction. Estimates of wind speed are unreliable. Chapter 6 will discuss the problems with changing instrumentation over the past century; it is, in any case, difficult to obtain a representative open ocean wind speed on board ship. Should the wind at 10 m be used, as is standard, or conditions closer to the sea-surface where the actual exchange processes are occurring? Estimates of sea surface temperature have an instrumental bias (see §6.4.1), while air temperature measurements are subject to distortion by ship heating. Another poorly parameterised variable is the cloud cover. The radiation formulae, such as (2.2), use the proportion of sky covered by cloud, as this is what is commonly measured. However, the radiational properties of different thicknesses, and heights, of clouds vary considerably. The properties of ice clouds will also differ strongly from those of water clouds.

This short discussion shows that significant uncertainties will remain in estimates of the heat exchange between ocean and atmosphere for some time. The climatic importance of this flux, however, means that considerable international effort is being put into improving satellite and surface observations, and also models, both theoretical and numerical. These research programmes include WOCE (World Ocean Circulation Experiment), JGOFS (Joint Global Ocean Flux Study), IGBP (International Geosphere–Biosphere Programme), and TOGA (Tropical Ocean, Global Atmosphere programme). Most of these are coordinated by the WCRP (World Climate Research Programme) of the World Meteorological Organisation.

2.4 Oceanic forcing by air–sea exchange of moisture and heat

2.4.1 *Moisture exchange*

We have repeatedly seen the climatic importance of latent heat, transferred to the atmosphere through evaporation. Evaporation, and its converse, precipitation, also help to drive the ocean circulation by creating horizontal density gradients. Evaporation, by removing water and concentrating the

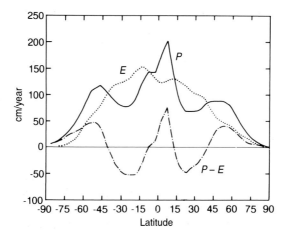

dissolved salts, increases salinity, and hence density. Precipitation, by adding water, reduces the salinity, and therefore density.

Density gradients produced in this way are less important to the regional ocean circulation than wind-induced flow, as will be seen in §2.10. Nevertheless, there are locations where these processes, evaporation especially, are the key to understanding the circulation. In §1.3.2 we saw how the Mediterranean Sea influences circulation in the North Atlantic ultimately because of its extreme evaporation. Similarly, large evaporation rates also occur in the Red Sea and the Persian Gulf.

The degree of excess of evaporation over precipitation is dependent on the atmospheric circulation. Its variation with latitude is shown in Fig. 2.9. Peaks in evaporation occur in the sub-tropical high pressure belts, while excess rainfall occurs in the mid-latitude westerly wind belts and the ITCZ (Inter-tropical Convergence Zone). A longitudinal variation also exists. The Atlantic is about 5% saltier than the other oceans because of higher evaporation relative to precipitation, *P*. This difference drives the present global thermohaline circulation (Fig. 1.14). Change to the longitudinal distribution of *E–P* may be responsible for the dramatic alterations to this circulation pattern seen in the last 20 000 years (Chapter 6).

2.4.2   Heat exchange

Section 2.3 may convey the impression that the ocean heats or cools the atmosphere but is not affected in return because of the high thermal inertia of the ocean. On the global scale this is not the case – the oceanic and atmospheric circulations are strongly coupled, as we will see in Chapter 5. However, to a large degree the oceanic coupling to the atmosphere for basin-scale processes is through the wind – *momentum transfer* – rather than heat transfer (§2.7).

Local heating and cooling of the ocean by the atmosphere is, nonetheless, important for climate. The seasonal change to the supply of thermal energy to the upper ocean away from the equatorial regions determines the characteristics of its *mixed layer*. This is the zone of the ocean immediately

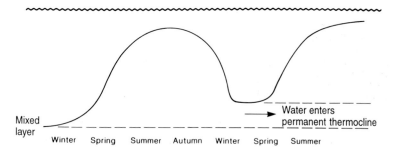

Fig. 2.10. A schematic diagram of the variation of mixed layer depth through the year. Note that each winter's maximum depth is not the same. In the example illustrated the second winter sees some water permanently lost from the influence of the atmosphere; if a future winter mixes back to greater depths the water then entrained into the mixed layer is likely to be of different origin.

below the air–sea interface, analogous to the atmospheric planetary boundary layer, where properties tend to be well mixed. In mid-summer the heating of the surface produces a thin layer of less dense water. This is well mixed by the wind but the mixing region is sealed by the narrow zone of rapid temperature (and therefore density) change at the base of the mixed layer. Cooling of this layer in the autumn reduces the density contrast at its base. This eventually leads to the mixed layer merging with the region below, or, in extreme circumstances, *overturning* to mix with a substantial extent of the water column. The warming of the surface in spring leads to the mixed layer becoming less dense. Rapid warming will lead to the formation of a new, shallow, mixed layer in the upper part of the winter layer. This formation of the summer mixed layer can occur over a few days given a warm, calm period of weather. The seasonal cycle in the mixed layer is illustrated schematically in Fig. 2.10. In Chapter 4 we will see that this cycle is of fundamental importance for biological processes in the upper ocean, controlling the variation of primary productivity through the year, and the supply of nutrients to the euphotic zone.

Locations where the mixed layer mixes only with water immediately below it can appear to possess two regions of rapid temperature change, as shown in Fig. 2.11. The deeper one, the *permanent thermocline*, corresponds to the depth of maximum winter mixing. The shallower zone, the *seasonal thermocline*, only appears in the summer and corresponds to the minimum summer mixing. The water between the two thermoclines in the summer, in being advected to another region, or because of variability in the severity of successive winters, can become detached from the region of seasonal surface influence to become a new sub-surface water mass in the ocean.

Fig. 2.12 shows the typical depth of winter mixing over the North Atlantic. In northern latitudes this is several hundred metres and in the Norwegian–Greenland Sea it can correspond to complete vertical mixing of the water column. In §1.3.2 we saw how this mixing was an integral part of the global circulation through the formation of North Atlantic Deep Water in the Norwegian and Greenland Seas (see Figs. 1.14 and 1.16). This type of extreme winter mixing, which also occurs in the Gulf of Lyon in the western Mediterranean, is due to pronounced cooling events in the atmosphere. Strong winds assist the overturning but the reduction of density contrasts by surface cooling is a necessary precursor. The combination of these two driving forces of the mixing means that the horizontal extent of overturned regions is likely to be restricted. Few observations of

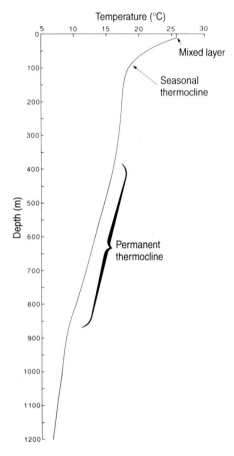

Fig. 2.11. The vertical profile of temperature at 33° 10.7′ N, 43° 12.5′ W in the North Atlantic on 7 August, 1983. The mixed layer is very shallow – less than 20m in depth. The seasonal thermocline extends from the base of the mixed layer to 100m, while the permanent thermocline is found from 400 – 900 m depth.

actual overturning events have been made but they suggest that regions less than a score of kilometres in diameter may be involved.

2.5 Basic forces within the atmosphere and ocean

There are five basic forces underlying the circulation of the atmosphere and the ocean: (i) gravity; (ii) pressure gradients generated by density or mass differences; (iii) drag, or the momentum gain or loss across the air/sea, air/land or land/sea interface; (iv) the *Coriolis force* due to the intrinsic rotation of the Earth; and (v) tidal motion due to the astronomical influence of the Moon and the Sun. Gravity and the pressure gradient force provide the basic balance in the vertical but horizontal motions in the atmosphere and ocean are driven differently. In the atmosphere the pressure gradient force and the Coriolis force strongly dominate, with surface drag acting to modify their balance, and tidal forces being almost irrelevant (discussion of drag in the atmosphere will be left to §2.7). All five forces can be involved in horizontal motion in the ocean. Local balance is provided by momentum transfer from the wind and the Coriolis force (the *Ekman spiral*). On larger scales density differences and the interaction of wind, sea and coastal boundaries produce pressure gradients, and also height differences. Tidal

Fig. 2.12. The average mixed layer depth over the North Atlantic Ocean at the time of maximum mixing (March). The depth of the mixed layer is determined by the depth at which the temperature has dropped by 0.5°C from the surface temperature. [Fig. 28 of Woods (1984), using data from Levitus (1982) ocean atlas. Reprinted with permission of J. D. Woods.]

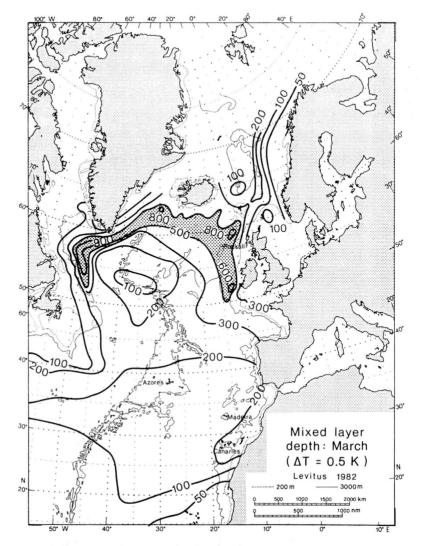

motions occur with regular periodicities to provide a strong mixing influence, and are often the predominant driving force in coastal zones (see §2.6). Motion resulting from all these influences is then affected by the Coriolis force. Detailed discussion of these motions in the sea will be considered in §§2.8–2.10, while some particular features of cloud physics and the marine atmospheric circulation will be considered in §2.11. A basic description of the key common processes will, however, be given here.

2.5.1 *Hydrostatic balance*

The atmosphere and ocean are generally close to being in *hydrostatic balance*. In the extreme this means that no vertical motion occurs because the force of gravity acting downwards is exactly balanced by the pressure gradient force acting upwards, away from the high pressures of the lower atmosphere or deep ocean generated by the weight of overlying air or water respectively. This balance is expressed by the hydrostatic equation:

$$dP = -\rho g dz \tag{2.7}$$

where dP is the change in pressure, P, over a very small, *differential*, height change dz, ρ is the density of the appropriate fluid, g is the gravitational acceleration and z is the vertical coordinate, which is positive upward. The ocean is essentially incompressible; a change of 10m in depth changes the pressure by about 10^5 Nm^{-2}, or the equivalent of one atmosphere. The density of the compressible atmosphere changes exponentially with height; near sea level a change of 10m in height changes the pressure by about 1 millibar (10^2 Nm^{-2}).

The hydrostatic approximation is not always obeyed in the atmosphere or ocean. During localised vigorous convection, such as in a cumulus cloud, deep water formation or fast flows down steep slopes, the balance is broken. However, even in situations where there is an element of large-scale vertical motion, such as in atmospheric frontal systems or the oceanic thermohaline circulation, hydrostatic balance is maintained locally. It is sometimes useful for such processes to be represented by a coordinate system whose vertical coordinate changes height slightly with position, following a pressure or density surface. In the atmosphere such a system is known as *sigma coordinates* while in the ocean these are *isopycnals*.

2.5.2   The Coriolis force

The motion induced by the pressure gradient force is affected by a number of other processes to produce the full, complex, three dimensional, nature of the circulation of the ocean and atmosphere. Foremost amongst these is the Coriolis force, a consequence of motion occurring on the surface of a rotating body, the Earth. The Coriolis force is often called a *pseudo-force* because if we were observing the Earth from space the effects ascribed to the Coriolis force could be completely explained by rotational dynamics. However, as the climate system is observed from the rotating Earth itself the introduction of such a pseudo-force makes it easier to describe the effects of the rotation.

To understand how the rotation of the Earth affects the motion of a moving body on its surface consider the following experiment. Take a disc and a water-soluble felt pen. Placing the pen at the edge of the disc lightly move it directly towards the centre whilst slowly rotating the disc anti-clockwise underneath. When you take your pen off the disc you will see that the path traced was a curve, that is, the pen acted as if it was 'deflected' to the right of its path. This occurred because the disc was spinning underneath the moving pen (Fig. 2.13(a)).

Now allow the pen to move in the direction of rotation around the disc. The pen will again trace a path curving to the right, as shown in Fig. 2.13(b). This time the cause is the desire for the pen to continue in a straight line, rather than be swept around in a circle. Once the pen begins to move

Fig. 2.13. Illustration of the Coriolis effect on a flat disc, rotating counter-clockwise. (a) The path drawn by the pen over the surface of the disc as it is moved directly towards the centre of rotation. (b) The path drawn by the pen as it is moved at right angles to a radius of the disc.

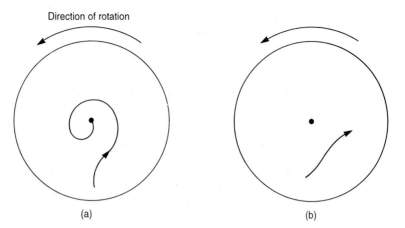

Direction of rotation

(a) (b)

outwards into a faster moving environment its velocity becomes less than the underlying rotation of the disc and it appears to move even further to the right, slowing down relative to the spinning disc.

Thus, for a disc spinning anti-clockwise a body moving on it will experience an apparent force to the right – the *Coriolis Force*, named after the French nineteenth century scientist who first correctly described the phenomenon. An anti-clockwise spin is experienced by objects moving on the northern hemisphere of the Earth; the southern hemisphere spins clockwise, seen from space below the South Pole, and so moving objects in this hemisphere experience an apparent force to the left.

A moment's thought will show the possibility of your pen ending where it started if the rotation of the disc continues for long enough. If an object is set in motion, and there are only very weak forces then affecting it other than the Coriolis force, it will continue to be deflected to the right (in the northern hemisphere) until it arrives back at its starting point. The Coriolis force is acting as a centripetal force so that

$$\frac{mu^2}{r} = mfu \tag{2.8}$$

where m is the mass of the object, r is the radius of the motion, f is the Coriolis parameter quantifying the effect of the Coriolis force ($f = 2\omega\sin\theta$, where ω is the rotation rate of the Earth and θ is the latitude at which the object is moving; see (2.10)) and u is the object's tangential velocity. This is known as an *inertial* oscillation. The angular velocity of the circular motion is thus f so that the period is $2\pi/f$. The period therefore depends on latitude, but is 12–24 hours in mid-latitudes. Such oscillations tend to occur in the ocean after sudden changes in the wind.

The Earth's Coriolis force is, however, acting on a sphere rather than a flat disc. We can see the origin of the Coriolis force on the sphere using some simple rotational dynamics.

Consider a ship steaming in the northern hemisphere, as shown in Fig. 2.14. As the ship steams north the distance from the ship to the Earth's axis

Fig. 2.14. Illustration of the
conservation of angular
momentum on a spinning
Earth. As the ship steams
north its distance from the
rotation axis decreases
causing it to move
eastwards, thus gaining an
eastward speed v, adding to
the underlying rotation
speed ω, and so spinning
faster (see equation (2.9)).

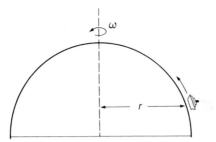

of rotation decreases. The angular momentum, I, of the ship is decreased also, as:

$$I = rv \tag{2.9}$$

where r is the radius from the axis of rotation and v is the rotational velocity (the local speed of the Earth in this case). In the absence of other forces angular momentum is conserved by a moving body. This can be illustrated by sitting in a revolving chair and spinning yourself around. If you hug your arms to your body, you decrease the radius of your *centre of mass* from the axis of rotation and consequently increase your speed, so that I remains constant. The northward-moving ship is in exactly the same situation: it needs to accelerate. It does this by turning to the east, so that its rotational speed becomes greater than that of the Earth, which is also rotating towards the east. Naturally, the opposite happens if the ship moves south; now it needs to decelerate and so turns to the west. In the southern hemisphere the opposite must happen.

The Coriolis force is also observed to act on objects moving east or west. The reason for this is not as easily seen as for the case of north–south, or meridional, motion, but again relies on rotational dynamics. When an object rotates it is kept in motion by a centripetal force, as shown in Fig. 2.15. This is directed at right angles to, and towards, the rotation axis. The object, however, has a velocity tangential to the radius, and so the force acts to pull the object towards the axis. As the object moves it is still subject to the centripetal force, but from a different direction. Continual adjustment to the direction of travel of the object maintains it in a circular path. The magnitude of this force can be shown to be mv^2/r or $m\omega^2 r$, where m is the mass of the object and ω is the rotation rate.

To an observer sitting on the rotating object (or on a stationary ship at sea) there appears to be no motion. For this observer, therefore, there must be another pseudo-force balancing the centripetal force. This is known as the *centrifugal force*, which acts in the direction of the radius, but away from the axis, as shown in Fig. 2.15. On the surface of the Earth this force can be expressed as the result of two right-angled components: one tangential to the surface, and a vertical component. When the object is at rest this tangential force, $m\omega^2 r\sin\theta - \theta$ is the latitude – does not induce any motion (as the Earth itself has distorted as a response to a balance between gravity and centripetal force, making the equatorial radius greater than the polar

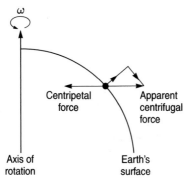

by 21 km). However, if the ship is moving to the east at speed u, it is effectively rotating faster than the Earth. Its centrifugal force is then:

$$m\left(\omega + \frac{u}{r}\right)^2 = \frac{m\omega^2}{r} + \frac{mu^2}{r} + 2m\omega u \qquad (2.10)$$

The first term on the right of (2.10) expresses the basic centrifugal balance of the Earth, while the second is much smaller in magnitude than the third. The horizontal component of this third term, $2\omega u \sin\theta$, known as the Coriolis parameter, f, gives the Coriolis force on a unit mass under zonal motion. At the equator there is no force, but away from the equator an object moving eastwards is pushed towards the equator, while one moving west is pushed poleward.

2.5.3 Geostrophy

The basic horizontal driving force in both the ocean and atmosphere is the pressure gradient force. The principal balance observed is with the Coriolis force; this is called *geostrophy*. However a horizontal difference in pressure is established a flow will be generated from high to low pressure. The Coriolis force acts to deflect this to the right (in the northern hemisphere) until the pressure gradient and Coriolis forces are equal and opposite. Steady flow proceeds, at right angles to the driving pressure gradient (Fig. 2.16). If the pressure gradient is known then this steady velocity can be calculated from the balance equations:

$$\frac{1}{\rho}\frac{\partial P}{\partial x} = fv$$
$$\frac{1}{\rho}\frac{\partial P}{\partial y} = -fu \qquad (2.11)$$

where (u,v) is the horizontal velocity in the (x,y), or (east,north), direction and, for example, $\partial P/\partial x$ is the change in pressure per unit distance in the x direction (a *partial derivative*).

In the atmosphere it is simple to measure pressure gradients, particularly at the surface, as pressure is one of the commonly observed quantities. In the ocean, however, pressure is often used as an approximate proxy for

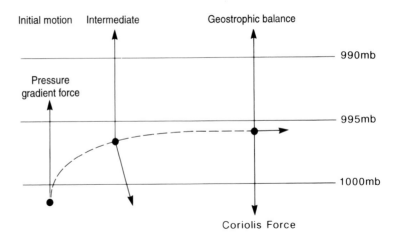

depth (because the pressure is mostly made up purely from the weight of overlying water) and the small horizontal gradients in pressure present have to be inferred from horizontal and vertical changes in density. This requires measurement of the variation of both temperature and salinity with depth. Even with such data it is rarely possible to obtain an absolute calculation of geostrophic velocity because there are slight gradients of the sea surface which need to be known, but are barely measurable even by the most advanced satellite instruments. Geostrophic velocities in the ocean are therefore usually calculated assuming some level of no motion, typically deep in the water column. This can introduce considerable error.

2.6   Tidal forces and their influence

Tidal motion in the ocean and atmosphere occurs due to gravitational interaction between the Earth and the other bodies in the solar system, principally the Moon and the Sun. The impact of the interaction of the Moon and Earth on a fluid envelope around the Earth is shown in Fig. 2.17. The two planets rotate about a common centre of mass, which, due to the greater mass of the Earth, is some 1700 km beneath the Earth's surface. A gradient of gravitational force pulling the fluid towards the Moon exists across the Earth because the force on an individual fluid parcel is proportional to the inverse square of the distance to the Moon. A second force acts due to the rotation of the Earth about the Moon–Earth centre of mass. This is a centrifugal force, directed away from the Moon everywhere over the Earth (see Fig. (2.17)). The combination of these two forces on the fluid envelope is to create bulges of fluid on the portions of the globe facing the Moon and opposite this, as shown in Fig. 2.17. The rotation of the Earth about its axis every 24 hours means that, for a given point on the Earth's surface, two peaks in sea level or surface atmospheric pressure pass over each day. This can be seen in the atmospheric pressure trace shown in Fig. 2.18, although it should be noted that the atmospheric tide is due to

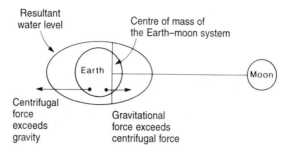

Fig. 2.17. The basic forces causing the lunar tide. The Earth–Moon system rotates about a centre-of-mass. On the Moon side of Earth the influence of the Moon's gravitational force is greater than the centrifugal force felt by the Earth due to this rotation (which should not be confused with the daily rotation of the Earth). On the side of the Earth away from the Moon the Moon's gravitational field is weaker while the centrifugal force is the same. The net result is a bulge in the water surface on both sides of the Earth.

pressure changes from the diurnal heating cycle of stratospheric ozone rather than gravitational forces.

Tidal motion shows rather more complexity than this simple model suggests. The *declination* of the Moon leads to unequal amplitudes of the two daily peaks, the *superposition* of the tidal effects of the Sun and Moon leads to bi-monthly modulation of tidal amplitudes, and the shape and depth of ocean basins results in oceanic tidal flow around basins rather than around the entire Earth.

The effect of the tides on the circulation and mixing of the atmosphere is essentially non-existent. By contrast, the tidal impact on the oceans can be substantial, particularly in coastal waters. The amplitude, phase and depth variation of the flow in such regions can be dominated by the tidal motion (see Fig. 2.19). Mixing of the ocean in regions of vigorous tidal currents can also be significant. This can be particularly true in regions of strong stratification and pronounced tidal amplitude because the *internal* tide (between layers of differing densities) can be of much greater amplitude than the surface effect.

For much of the discussion in the rest of this book, however, tidal effects are likely to be of limited importance. Exceptions to this will occur for processes where coastal effects are significant in the oceanic budget and these will be highlighted when encountered.

2.7 Momentum transfer and drag

The main driving force of the ocean surface circulation is the wind. Blowing over the surface, the wind exerts a stress on the ocean, transferring momentum from the air to the water. This is then, in ways to be discussed in the next three sections, mixed into the fluid, producing motion of various scales.

At the same time as the wind pushes the sea, the water, in extracting energy from the wind, is acting as a drag force on the atmosphere. This slows the wind near the surface, which acts as a drag higher in the air. It also turns the near-surface wind away from its geostrophic direction to give *cross-isobaric* flow towards the lower pressure (Fig. 2.20). The direct influence of surface drag in the atmosphere is to about a kilometre in height, although it can be much less in light winds. The vertical profile of horizontal wind speed in this boundary layer is logarithmic in many circumstances. As

Fig. 2.18. A barograph
trace recorded in Norwich,
England, from 0800 GMT
on 30 October, 1993, to
1700 GMT on 1
November, 1993.
Superimposed on the
gradual decline in pressure
over this period is a twice
daily oscillation, the peaks
of which are indicated by
arrows. This oscillation, of
about 1 mb in amplitdue
(the vertical scale is in
millibars), is due to the
atmospheric tide. Trace
courtesy of John Green.

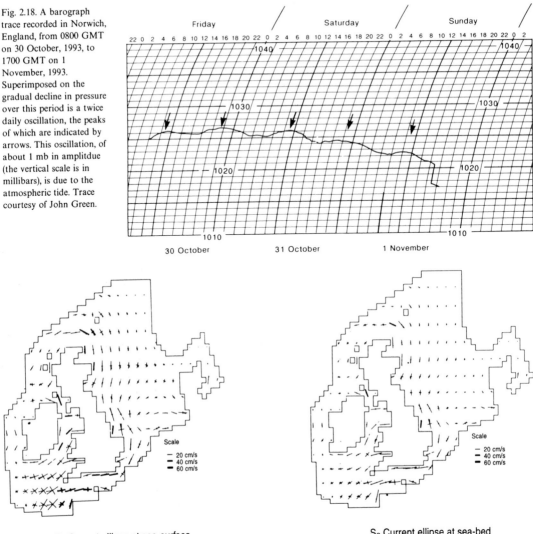

Fig. 2.19. Magnitude of the S_2 oceanic tide (the semi-diurnal solar tide), calculated from a numerical model of the tides of the European shelf. The tide at the sea surface and sea bed is shown in the form of tidal ellipses. Following a vector around the ellipse formed from the two right-angled axes at a given location describes the evolution of the tidal current through a full tidal cycle. Along the east coast of Scotland, for example, the tide essentially completely reverses direction between ebb and flood tide. However, southwest of England there is a strong circular element to the tidal variation. [Fig. 7 of Davies (1986). Reproduced with permission from Kluwer Academic Publishers.]

the wind speed above the boundary layer is independent of the ground immediately below it, the steepness of the natural logarithm depends on the roughness of the underlying surface and the turbulence of the atmosphere at the surface:

$$u = \frac{u_*}{k}\ln\left(\frac{z}{z_o}\right) \tag{2.12}$$

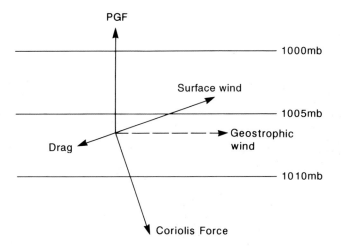

Fig. 2.20. The balance of
forces in the atmospheric
boundary layer. The
surface wind is deflected in
the direction of the
pressure gradient force
(PGF) due to the retarding
effect of drag, which also
reduces the size of the
Coriolis force, as this is
proportional to speed.
Note that the Coriolis
force must always be at
right angles to the velocity.

In equation (2.12) z_0 is the *roughness length* of the surface, $k \sim 0.4$, and u_* is the *friction velocity*. The roughness length gives the height at which the surface drag brings the atmosphere to rest, and can be very crudely thought of as the typical height of a perturbation to a flat surface (this is not quite correct, as z_0 is usually smaller than this, but it gives a first approximation). The friction velocity is a measure of atmospheric turbulence near the ground. In a strongly turbulent atmosphere the horizontal and vertical velocities will be similar in magnitude: in this extreme, u_* is large and the horizontal velocity will change rapidly with height. In a nearly *laminar* atmosphere the horizontal velocity is much greater than the vertical velocity: u_* is small and the horizontal velocity will change slowly with height.

Equation (2.12) gives the vertical profile of horizontal velocity over any surface. The roughness length and friction velocity are often determined by taking two measurements of velocity at different heights and solving the two simultaneous equations that (2.12) then provides. Over the sea the friction velocity is about a twentieth of the 10 m wind speed; z_0 varies from 10^{-6} m at wind speeds of 3 ms^{-1} to 10^{-4} m at 10 ms^{-1}. Over the land z_0 is a few centimetres above grass, and metres in a city. Land surfaces, having greater roughness which induces turbulence, exert considerably greater drag on the atmosphere than does the sea.

The logarithmic profile gives the wind near the surface a vertical gradient, or *shear*. This provides the mechanism for momentum to be transferred down towards the surface. A shear flow is not stable because small disturbances tend to grow, making the fluid turbulent. This turbulence, consisting of small eddies which provide the gustiness of the wind, acts to alter the shear. Faster moving parcels of air from above tend to move downward, while slower moving eddies from below move upwards. This produces a net transfer of momentum downwards. This momentum is then captured by the surface as its drag force operates. The stress is clearly related to the mean wind speed as well as shear. Conventionally, it is empirically evaluated relative to the 10 m wind speed, u.

The stress, τ (measured in kgm^{-1}s^{-2}), acting on the sea surface is given by

$$\tau = C_D \rho_a u^2 \tag{2.13}$$

where C_D, the drag coefficient, is a non-dimensional coefficient. The drag coefficient is an important parameter, as it determines the proportion of the atmospheric boundary layer momentum which is converted into ocean currents. As a first approximation C_D can be taken to be a constant: $C_D \sim 1.5 \times 10^{-3}$. This is reasonable for wind speeds of moderate strength, between 5 and 15 ms^{-1}. However, C_D will, in practice, depend on the type of logarithmic profile, and, hence, the roughness length and wind speed. Many experiments have been performed to find a relationship between the wind speed and C_D; most produce a linear fit, such as

$$C_D = (0.61 + 0.63u) \times 10^{-3} \tag{2.14}$$

There is much scatter in the experimental results, and such equations as (2.14) must be regarded as approximations only, especially at low and high wind speeds. In both light and strong winds the degree of gustiness, for example, will be very important for downward transmission of momentum; current study of the provision of stress to the sea surface is concentrating on relating this stress to such atmospheric variability. Note that the magnitude of C_D shows that the transfer of momentum from atmosphere to ocean is very inefficient; the surface ocean currents are thus typically only 1–2% of the wind speed.

2.8   Waves, the production of aerosols and condensation nuclei

The moving atmosphere produces a stress on the sea surface. The ocean is a fluid and exertion of a stress upon it will result in motion within the fluid. This takes many forms. To the atmospheric observer the most obvious of these is the production of surface waves. The type of waves, or the *sea state*, is a clear function of the wind speed; light zephyrs give tiny ripples, great storms produce towering mountains of foaming water. The sea state is of vital importance for the air–sea exchange of atmospheric gases that play important roles in oceanic chemistry and biology of climatic relevance, such as CO_2 or O_2 (see Chapters 3 and 4). A violent sea full of breakers strongly enhances this exchange through bubble generation. The exchange process, and the physical break-up of waves, is also important for the atmosphere as oceanic particles and gases are significant sources of cloud *condensation nuclei*. Thus an understanding of surface wave processes is a necessary prerequisite to an understanding of ocean–atmosphere climatic interaction.

2.8.1   *Wave formation and characteristics*

Surface waves occur due to subtle interactions between small-scale pressure fields in the boundary layers of the atmosphere and ocean. The full, dynamic, wave field that we see on the ocean surface springs from these interactions, but is also affected by interactions within the sea between waves of different frequency.

Fig. 2.21. Photograph of wave breaking along the Dorset coast near Golden Cap on 11 August, 1985. A strong onshore wind, combined with a rapidly shoaling beach at a stream inlet, has led to vigorous surf. Note the size of the breaking waves in relation to the adults standing on the beach.

If surface waves build up in size they can reach a peaked shape which becomes gravitationally unstable because the leading edge of the wave crest leans over the front of the wave itself. The wave then breaks as this crest collapses. Momentum is released, taken by the waves from the wind, to the body of the ocean.

Breaking can be caused in several ways. Strong winds may provide the energy to amplify waves to this extent. Combinations of waves of smaller size from different source regions may locally reinforce to produce sufficiently large waves for breaking to occur; this is known as *constructive interference*. Both of these modes of breaking will be observed at sea; the latter tends to produce localised patches of wave breaking. The major source of breaking waves observed from land is the amplification caused by the waves travelling over a shoaling beach. This can be extremely dramatic, as shown in Fig. 2.21. Coastal wave breaking occurs because the shoaling sea floor is squeezing the wave energy into a smaller depth range. To conserve energy (though, of course some is being lost due to friction at the bottom) the wave height grows. As a rough guide, wave breaking occurs as the height of a wave approaches one seventh of its wavelength.

The spectrum, or combination of sets, of waves depends on the wind conditions, both locally and at distance. There tends to be a dominant frequency, with little energy in waves with longer period (or smaller frequency) but considerable energy at shorter periods. Stronger winds produce a sharper peak in energy, at a lower frequency, as shown in Fig. 2.22. Each set of waves of differing frequency moves at a different speed (see (2.15) below), radiating energy in all directions from its source. This has the effect of dispersing energy from the generation area. Such dispersion is particularly noticeable some distance from a storm, as the waves radiated by the storm-induced wave field arrive at a coast at different times depending on their frequency. The *swell* produced by the storm will

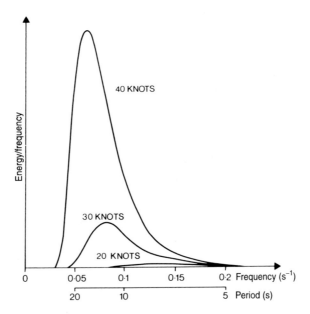

Fig. 2.22. Variation of the spectrum of normalised wave energy with frequency for three different wind speeds (1 knot = 0.51 ms^{-1}). [Reproduced, with permission, from J. G. Harvey, *Atmosphere and ocean: our fluid environment* (London: Artemis Press, 1985), p. 77, Fig. 9.4]

therefore appear very regular, in contrast to the continually changing wave field produced locally.

We noted earlier that surface waves attain greater amplitude on approaching the shore. This illustrates two important facts about surface waves: they carry energy, and their influence extends some distance beneath the surface. Surface waves transmit energy but not matter. The particles which appear to be carried along with the wave actually move in circular orbits. The appearance of the surface wave is due to the phase of the orbit of the surface particles. The direction of travel is prescribed by the orbital orientation. As the wave approaches the coast, and the sea floor rises, the bottom begins to interact with the wave, normally where the water depth is less than half the wave's wavelength. The orbits become elliptical in shape as this occurs. In general the phase speed, c, of the waves is

$$c = \lambda v \tag{2.15}$$

where v is the frequency and λ is the wavelength. In shallow water, where the wave is affected by the bottom (and where depth is typically less than a twentieth of a wave's wavelength), the waves become independent of frequency, and lose their open ocean dispersive character, with

$$c = \sqrt{gH} \tag{2.16}$$

H being the depth of water.

2.8.2   *Breaking waves and marine aerosols*

Waves break at sea, and on impact with the shore. Fig. 2.21 illustrates important facets of the wave breaking process. It shows large numbers of

Fig. 2.23. Schematic of bubble bursting at the sea surface. The lower picture shows a bubble rising through the water which has just reached the air–sea interface. Milliseconds later, in the top picture, the bubble has burst, with a jet of water, caused by the pressure differences between the bubble and its environment, throwing water droplets into the air.

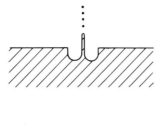

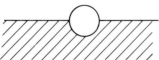

whitecaps, and spray thrown liberally into the air. Whitecaps consist of strongly oxygenated water; breaking waves are an important pathway for atmospheric gases to enter the sea. The spray shows that the breaking waves throw sea water into the air. Some of this is evaporated before falling back into the ocean; more, with its dissolved salts, is carried away by turbulence to contribute to the atmospheric collection of condensation nuclei.

A typical vertical profile of the concentration of dissolved oxygen is shown in Fig. 3.9. The surface maximum is accentuated by breaking waves injecting additional gas into the water. Regions of strong winds, and many breakers, will therefore have enhanced levels of dissolved oxygen in their surface waters. We shall return to the mechanisms of gas transfer across the air–sea interface in Chapter 3.

Many of the bubbles that are injected into the water as a wave breaks, however, rise back to the surface before their gases dissolve. As the bubble breaks through the surface its pressure, being higher than the atmosphere's, causes it to explode, with a jet of water erupting from the base of the bubble, as illustrated in Fig. 2.23. This jet breaks into a collection of droplets, some of which fall back into the water, but some of which are whisked away by the wind. The splashes resulting from the impact of the remnant jet droplets will also add further water droplets, and salt particles, to the air.

Bubbles would seem to be directly linked to wave breaking, and indeed experiments at low wind speeds, using sonar to trace bubble flocks, show discrete bubble events associated with breaking waves. This can be seen when travelling by aeroplane as a burst of white water appearing where, a second or two earlier, a wave has just broken. At higher wind speeds, sufficient wave breaking occurs for large areas to have considerable bubble concentrations to depths of 10 m or more. A background population of bubbles exists, however, even in the absence of breaking. This appears to be because of the longevity of some bubbles within the water. Addition of salt particles to the atmosphere will consequently occur continuously, although the process will be significantly enhanced during periods of strong winds.

The salt particles injected into the atmosphere through bubble bursting, splashes from breaking waves and water droplets torn by the wind from the sea surface provide the majority, by mass, of the condensation nuclei over

Table 2.2. *Sources for condensation nuclei over sea and land*

		Concentration(μgm^{-3})	
Substance	Sources	Land	Ocean
SO_4^{2-} Sulphate	Oxidation of SO_2 Combustion of fuels, forest fires, volcanoes, marine biology, sea spray, airborne soil particles	1 (10 in polluted air)	0.1–1.0
NH_4^+ Ammonium	Combustion, decay of organic matter in soils	0.1–1.0 (10 in polluted air)	0.01
NO_3^- Nitrate	Industrial processes, automobiles, sea spray	<1 (1 in polluted air)	0.1
NaCl Sodium chloride	Sea-salt particles, some soil particles	<1	1–10
Organic carbon	Combustion, automobiles	1	0.1–1.0

maritime areas, and a significant proportion over land. Condensation nuclei are crucial to cloud formation and are considered next.

2.8.3   *Condensation nuclei*

In the previous references to cloud formation (§§ 1.7, 2.2.1) we have implied that clouds form due to condensation of atmospheric water vapour into droplets. This is correct, but if the condensation is unaided by some initiating mechanism then supersaturations of 400% can occur before the spontaneous appearance of droplets. The atmosphere, however, is never completely clean: there are always particles, or *aerosols*, present. These act as nuclei for condensation, allowing droplets to form with only very small supersaturations of less than 1%. Some nuclei are *hygroscopic*, which means that water molecules tend to stick to their surfaces at humidities below saturation, in some case as much as 15% below. Anyone who has left a salt cellar exposed to the air will know that salt strongly attracts atmospheric water vapour! Such particles are thus of major significance for aiding cloud formation and rainfall, particularly for air temperatures above freezing. Typical sources for cloud nuclei are shown in Table 2.2.

The formation of ice in clouds, characteristic of high clouds such as *cirrus*, or the upper reaches of vigorous cumulus clouds or the *cumulonimbus* associated with thunderstorms, is also dependent on aerosols. Drops of water of cloud droplet size – a few micrometres – will not spontaneously freeze until the temperature reaches $-40°C$. Ice condensation nuclei allow ice crystals to form at temperatures only a few degrees below $0°C$. Clay particles are better nuclei than sea salt for ice crystal formation, and hence snow production. Silver iodide is another material that aids ice formation; it was used extensively in the 1960s and 70s when cloud seeding experiments were performed in the (essentailly unrealised) hope of aiding rainfall initiation.

In the maritime environment sea salt is the major aerosol by mass; over land it is important but other particles have similar, or greater, concentrations in the lower atmosphere[4]. The fact that the oceans cover 70% of the surface area of the globe means that production of sea salt particles, as discussed in the previous section, is of considerable climatic impact. Alteration to marine tropospheric circulation, modifying the wave regime of the ocean, and therefore the distribution and quantity of wave breaking, may feedback on climate by ultimately influencing cloud formation, rainfall and the Earth's radiative properties. This will be pursued further in Chapters 3,4, 6 and 7.

2.9   The Ekman spiral and Langmuir circulation

The orbital motion of water induced by surface waves penetrates some metres beneath the sea surface. The direct effect of the wind, however, penetrates to a depth of scores of metres. This is because of a larger scale physical force balance than the microscale pressure anomalies causing wave motion. The resulting flow varies with depth in the Ekman spiral. Small-scale circulation cells driven by this spiral are known as Langmuir circulation cells.

2.9.1   *The Ekman spiral*

In the winter of 1893/4 Fridtjof Nansen's vessel, *Fram*, was stuck fast in the Arctic ice pack, 500 kilometres north of Russia. This was not due to poor seamanship, but a deliberate attempt, using a specially designed vessel, to undertake scientific study of the Arctic during the inhospitable winter. The ice voyage was also used to give a northerly launching point for Nansen and Johansen's unsuccessful attempt to reach the North Pole.

The *Fram* slowly drifted in a westerly direction, the ice being pushed by the prevailing near-surface ocean current. Nansen observed that the ice usually moved at an angle of 20–40° to the right of the wind direction. On his return to Norway he discussed this phenomenon with Ekman, who developed a theory to explain it.

We saw in §2.7 that the wind exerts a stress on the ocean surface. Orbital wave motion enables the upper ocean to act as a series of layers. The stress transmitted to the surface layer by the wind forces motion in the layer below, through the stress at the layer interface. This passes momentum to a further layer, so driving flow in a succession of layers, as shown in Fig. 2.24. Each layer dissipates energy through internal friction, allowing the wind stress to penetrate only so far into the ocean – this depth defines the *Ekman layer*. As each layer is moving it is subject to the Coriolis force, which will deflect its motion to the right in the Northern Hemisphere. If the wind has been blowing with the same strength and direction for some time, our observation point is far from land so that coastal effects do not modify the

---

[4] While sea salt is a major aerosol, sulphate and organic carbon form the majority of the actual particles. These, however, have much smaller radii. The formation, and importance, of these smaller cloud condensation nuclei will be considered in Chapters 4 and 7.

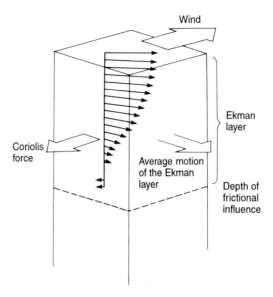

Fig. 2.24. Schematic of the effect of a steady wind blowing over the surface of the ocean. Over the Ekman layer successive layers of the water frictionally induce motion in the layer beneath, until no energy remains. Each layer's motion is subject to the Coriolis force; the net Coriolis force over the total column opposes the surface wind stress giving the Ekman transport of the layer – the net motion of the column as a whole – to the right of the wind in the Northern Hemisphere. [After Fig. 3.6 of Open University Course Team (1989). Adapted with permission of Butterworth–Heinemann Ltd.]

flow, and if the water is of uniform density then, for a steady flow, there must be a force balance between the wind stress and the net Coriolis force integrated over the Ekman layer. This is illustrated in Fig. 2.24. The resulting net flow over the Ekman layer is therefore at right angles to the forces – just as for the geostrophic velocity discussed in §2.5.3 – and to the right of the wind in the northern hemisphere. This net flow is known as the *Ekman transport*, and is a very important principle for understanding the surface ocean circulation, as we will see in §2.10.

Just as the wind exerts a stress on the water surface we saw in §2.7 that the ocean exerts a stress or drag on the atmosphere. The same physical argument can therefore be applied to the atmospheric boundary layer, even over land. An Ekman layer is thus present in the atmosphere, with the force balance acting in the opposite direction to give the atmospheric Ekman transport a backing to the left in the northern hemisphere. To balance the oceanic stress, the atmospheric Ekman mass transport will be the same as the oceanic. However, the volume transport will be much greater because of the large density difference between air and water. The atmospheric Ekman layer is thus significantly deeper than the oceanic layer, reaching perhaps a kilometre in height.

In Fig. 2.24 the surface current is shown directed at an angle of 45° to the wind. This direction was predicted by Ekman and is caused by the influence of viscosity near the surface, where there are strong vertical current shears. The surface current is actually aligned with the geostrophic wind velocity, but the friction of the air with the surface causes the surface air velocity to deviate, as discussed in §2.5. As Nansen observed, the surface current is rarely, if ever, at its theoretical orientation relative to the wind. The Ekman spiral is almost always concealed from view by the effect of interaction with coasts or the sea floor, and also by the currents driven by density differences within the ocean. The net Ekman transport is nevertheless generally observed as a vital feature of the upper ocean circulation.

Fig. 2.25. Net Ekman transport for sub-tropical oceans in the Northern Hemisphere. The anticyclonic surface winds (a) produce net inflow of water towards the centre of the sub-tropical high pressure centres. This drives convergence (b) and downwelling of water in the centre of the oceanic gyre, both raising the sea surface, and depressing the thermocline. [After Fig. 3.23b of Open University Course Team (1989). Adapted with permission of Butterworth–Heinemann Ltd.]

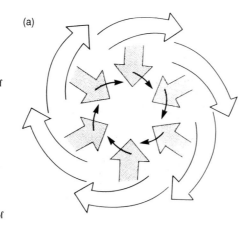

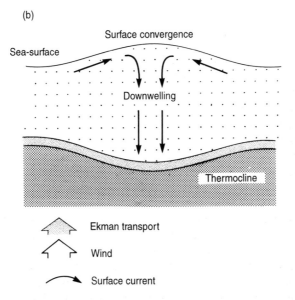

Fig. 1.6 showed the mean circulation of the lower atmosphere. If we regard this as steady over several months, so that the balance leading to the Ekman spiral can be established, the variation of wind speed and direction with position will create zones where Ekman transports from different directions will oppose. For instance, over the sub-tropical oceans the anticyclonic atmospheric circulation will induce Ekman transports towards the centre of the anticyclone, as shown in Fig. 2.25. This will produce convergence, with a doming of the sea surface and consequent sinking of water. This downwelling is known as *Ekman pumping* and the resulting vertical velocity can be shown to depend on the horizontal gradients in the wind stress. Although this vertical velocity is generally less than 50 cm/day it is the main contributor to vertical motion in the upper ocean. During strong storms or hurricanes, when winds change strikingly over small distances, Ekman pumping can be considerable. In Chapters 3 and 4 we will

Fig. 2.26. A photograph of
streak lines associated with
foam convergence along
the windrows of Langmuir
circulation cells. [Fig. 3.26b
of Open University Course
Team (1989). Reproduced
with permission of
Butterworth–Heinemann
Ltd.]

Fig. 2.26. A photograph of streak lines associated with foam convergence along the windrows of Langmuir circulation cells. [Fig. 3.26b of Open University Course Team (1989). Reproduced with permission of Butterworth–Heinemann Ltd.]

see the importance of vertical motion for chemical and biological processes of climatic relevance; §2.4.2 has already referred to this.

2.9.2 *Langmuir circulation*

If you have flown in an aircraft over the sea, or been able to observe a lake from surrounding mountains you will have probably observed parallel streaks of foamy water, 30–50m apart, extending hundreds of metres across the water. A photographic example is shown in Fig. 2.26. These are the surface signatures of regions of convergence within long circulation rolls aligned with the wind direction, as illustrated schematically in Fig. 2.27. This circulation system – called Langmuir circulation – was first studied extensively by Irving Langmuir in 1938, when he observed long rows of seaweed arranged parallel to the wind in the North Atlantic Ocean.

Various theories have been developed to explain the continuance of established rolls through reinforcement of the convergence by the wave field. The origin of the circulation cell is not well understood. It is believed to be the result of an instability in the Ekman spiral, producing variation in the vertical shear; hence the circulation penetrates to the bottom of the mixed layer. The atmosphere also shows such behaviour, although on a larger spatial and temporal scale. Analogous cloud streets, often seen in clear cold air following the passage of a cold front, require hours to form, rather than the 20–30 minutes for Langmuir circulation development.

2.10 Wind-driven circulation of the ocean

In the last three sections we have seen how the wind acts directly on the ocean and the local effect of ocean surface drag on the atmosphere. The momentum transferred to the water, plus the heat and fresh water, creates pressure gradients in the ocean leading to motion on a larger scale. The

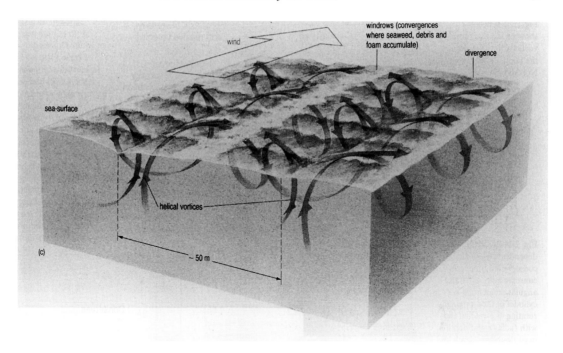

Fig. 2.27. Schematic of the circulation associated with Langmuir circulation cells. [Fig. 3.26c of Open University Course Team (1989). Reproduced with permission of Butterworth–Heinemann Ltd.]

larger scale impact of the ocean on the atmosphere is also through the creation of pressure gradients, but via surface heating or cooling. The Coriolis force then acts as well, to give geostrophic balance and maintain the general circulation in each fluid. In this section we will discuss these basin-scale flows in the ocean while in §2.11 and Chapter 5 we will discuss the large-scale impact of surface heating on the marine atmosphere.

2.10.1   The ocean gyres

The key to understanding ocean gyres is contained in the concept of angular momentum that we discussed in §2.5.2. Conservation of angular momentum led to the Coriolis force. The winds of the sub-tropical anticyclones provide the ocean with angular momentum. As these winds blow semi-permanently the oceanic sub-tropical gyres should gradually accelerate. They do not; the gyre circulation is stable. Therefore there must be a mechanism providing a source of oppositely rotating angular momentum to balance the wind's influence.

Meteorologists and oceanographers use the concept of *vorticity* to replace angular momentum because there are two ways a body of water can gain angular momentum, and vorticity combines them. As we saw in our discussion of the Coriolis force, a body of water will change its angular momentum relative to the Earth's axis when moving north or south. This type of angular momentum is called *planetary vorticity*. We have also seen several situations where water rotates about a local axis of rotation, for

the basin there would be an excess of negative vorticity leading to an acceleration of the rotation. For balance the positive vorticity given by friction in the west needs to be high. This is achieved by a strong western boundary current trapped in a narrow zone near the coast. The Gulf Stream, in the North Atlantic, is such a current.

These strong boundary currents are important components of the climate system as they transfer large quantities of warm water poleward very rapidly. The Gulf Stream, for example, has a mean flow of about 100×10^6 m^{3}s^{-1} If the Gulf Stream off the northern United States were only 2°C warmer than the global zonally averaged sea temperature for 40°N then this would represent a poleward energy flow of 10^{15} Js^{-1}, approximately 20% of the total northward heat flow in the atmosphere and ocean combined at these latitudes (this is actually an under-estimate of the local temperature anomaly in the Gulf Stream). The western boundary currents are as important in the ocean for poleward heat transport as mid-latitude depressions are in the atmosphere. It is easily seen that any change in gyre circulation has a significant impact on the climate system.

The boundary currents exhibit another feature, particularly after separating from the coast. The strong shear in flow across the current, and the thermal contrast between the warm sub-tropical waters equatorward and the cold waters poleward, leads to instability. The current meanders, occasionally leaving intrusions of warm or cold water isolated in the cold or warm regions respectively. This process assists the diffusion of heat polewards, but also has implications for biological productivity which we will discover in §2.10.6.

It takes a month or two for water to be transported in this boundary current from the tropics to the mid-latitudes, but numerical models suggest full circulation requires 15 years or more. Ekman transport suggests that during this circulation there is a net movement of water towards the centre of the gyre, as noted in §2.9.1. The accompanying doming of the surface, while only a few centimetres, maintains, and is maintained by, the geostrophic gyre circulation. The converging water will form a new water mass at intermediate depths in the ocean.

Poleward of the sub-tropical gyres the atmospheric circulation imparts positive relative vorticity to the ocean, as the surface winds swing from westerly to easterly. A sub-polar, cyclonic gyre would therefore be expected. Vorticity arguments imply that there should be strong western boundary currents in such a gyre as well. No such gyre exists in any of the southern hemisphere oceans, because of the circumpolar ocean link at 60°S (see §2.10.6). In the northern hemisphere there are weak gyres, heavily constricted by northern coasts, in both the Pacific and Atlantic Oceans. Their western boundary currents are consequently rather weak.

2.10.2 *Coastal upwelling*

The sub-tropical anticyclones produce equatorward winds along the eastern shores of the oceans, as shown in Fig. 1.6. The Ekman transport

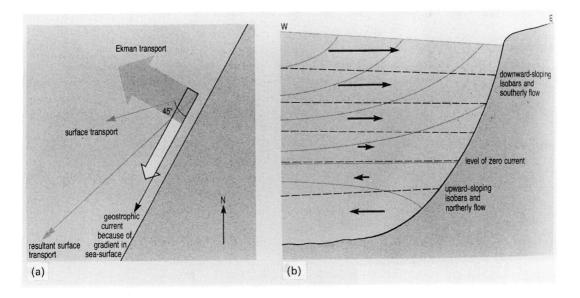

Fig. 2.31. Illustration of (a) the surface transport due to the off-shore Ekman transport in regions of coastal upwelling in the Northern Hemisphere, and (b) sub-surface structure and velocity field. Arrows are the pressure gradient force and light grey lines are isopycnals. [Fig. 4.30c and d of Open University Course Team (1989). Reproduced with permission of Butterworth–Heinemann Ltd.]

associated with these winds is off-shore. To balance this mass transport, cold water episodically upwells from below the surface, bringing high concentrations of nitrates and phosphates to the euphotic zone. These regions of coastal upwelling are therefore very biologically productive, as such chemicals are the basic food source for plankton (see Chapter 4). It is important to note that such regions are not found on western coasts of the ocean basins because of the asymmetry of the position of the anticyclones producing the wind forcing. The reasons for this will be pursued in Chapter 5.

The strongest upwelling occurs off the west coast of northern and southern Africa, and off the west coast of South America. The abundant plankton provide a base for the food chain in these waters, leading to rich fisheries. The upwelling zone off South America is particularly known for its anchovy fishing. This region is, however, subject to occasional periods of dramatic reduction in the upwelling as a component of the climatic disturbances associated with El Niño: a coupling of the ocean and atmosphere causing the climate of the tropical Pacific basin, and further afield, to be highly anomalous for periods of a year or more. El Niño will be discussed in detail in Chapter 5.

Upwelled water, being colder and denser than that moved westwards by Ekman transport, does not quite replace the volume of the lost water. A slope is therefore set up off-shore, as shown in Fig. 2.31. This slope creates a pressure gradient towards the shore, with a consequent geostrophic current aligned alongshore with the wind direction. The observed surface ocean current is therefore a combination of the off-shore Ekman transport and this alongshore geostrophic flow; the result is at an angle to the coast, as in Fig. 2.31.

Fig. 2.32. Infra-red image
from the geostationary
Meteostat satellite, taken at
1030 GMT on 11 April,
1995. Note the vigorous
convection over central
Africa and the set of
storms over the tropical
Atlantic which are part of
the ITCZ. Supplied by the
Department of
Meteorology at the
University of Edinburgh.

2.10.3   *The tropical surface circulation*

The wind systems of the two hemispheres meet in the tropics. We saw in §1.2 that the easterly winds on the equatorward side of the sub-tropical anticyclones converge in a semi-continuous line called the Inter-Tropical Convergence Zone, or ITCZ. The converging flow forces air to rise, creating the tendrils of massive thunderstorms seen in satellite photographs such as Fig. 2.32. This narrow band of storms moves in response to the seasons, tending to be in the summer hemisphere. Fig. 2.32 shows a clear dependence of the positioning of the ITCZ on the land/sea distribution, with heated land masses, such as central Africa, pulling the ITCZ poleward of its latitude over the neighbouring oceans.

The junction of the hemispheric Trade wind systems in the vicinity of the equator leads to interesting possibilities for the dynamical interaction of the ocean and atmosphere. This interaction is illustrated in Fig. 2.33. As the ITCZ tends to form in the summer hemisphere the Trade winds from the winter hemisphere must cross the equator. This circulation is subject to oppositely directed, diverging, Coriolis forces, on either side of the equator. The ocean circulation is also subject to these forces, and the resulting Ekman transport causes divergence of surface waters from the equator. A

Fig. 2.33. Schematic of the various currents in the tropical oceans and their connection to surface divergence or convergence and the wind field. [Fig. 5.1a from Open University Course Team (1989). Reproduced with permission of Butterworth–Heinemann Ltd.]

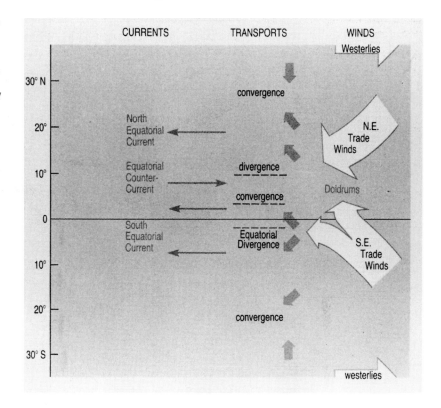

narrow strip of ocean along the equator will therefore need upwelling to compensate for this Ekman flow. This zone is clearly seen as an equatorial tongue of cold water in the sea-surface temperature pattern of Fig. 2.34, particularly in the Pacific Ocean. The horizontal surface circulation at the equator itself mirrors the westward wind stress, as the Coriolis force vanishes here. The steady winds can induce strong currents of more than $1\,\mathrm{ms}^{-1}$. The carbon exchange between ocean and atmosphere is also distinctly anomalous along the equator, as will be discussed in Chapter 3.

Poleward of the equator the main atmospheric, and oceanic, circulation consists of the equatorward side of the sub-tropical anticyclones, or gyres. The flow in both media is therefore predominantly westward. In the ocean this creates what are known as the North and South Equatorial Currents, shown in Fig. 1.15. This westward flow, coupled with the weakness of the Coriolis force very close to the equator, creates a gradient in the sea surface across the ocean basin. The resulting eastward pressure gradient opposes the westward flow, generated by the equatorial easterly winds, which acts to support the sea surface gradient. The upper ocean circulation pattern is balanced by water flowing below the surface eastwards, and towards the equator, at a depth of about 100 m. This converges to form the, eastward flowing, Equatorial Undercurrent. This can also reach high speeds of well over $1\,\mathrm{ms}^{-1}$, the strongest flow being tightly constrained near the equator in most circumstances.

The equatorial oceans are thus regions of strong gradients and strong

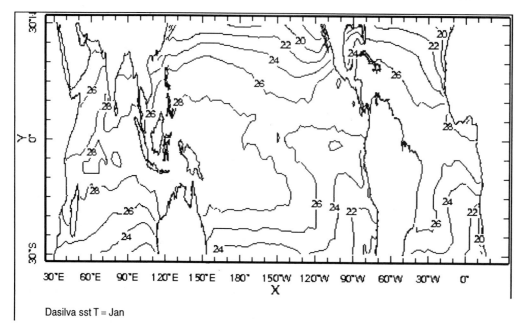

Dasilva sst T = Jan

Fig. 2.34. Average sea
surface temperature over
the tropical oceans during
January. Contour interval
is 2°C.

currents. Near the ITCZ itself yet another dynamic interaction occurs, as shown in Fig. 2.33 for a northern hemisphere summer. Here we find a region of light winds – the Doldrums – as the Trade wind systems mix. The Ekman transport produced by the summer hemisphere Trades, which have not crossed the equator, is away, and hence divergent, from the Doldrums. The winter hemisphere Trades, having crossed the equator, also produce an Ekman transport away from the equator. This transport, however, causes a convergence of water on the equatorward side of the Doldrums. The combination of this divergence and convergence creates a sea surface slope across the Doldrums. The resulting pressure gradient is poleward, producing a geostrophic flow to the east: the Equatorial Counter-Current.

The upper ocean dynamics illustrated schematically in Fig. 2.33 are theoretical in nature. In the central and eastern Pacific the ITCZ remains north of the equator throughout the year, while in the west the ITCZ breaks dramatically to the south in the southern summer. This will complicate the current pattern considerably in the west Pacific. The Atlantic sector, by contrast, is closer to the theoretical picture, with the ITCZ moving substantially between the seasons. The strength of the southern hemisphere Trades here can vary substantially with latitude at certain seasons, giving rise to a weak counter- current south of the equator.

The Indian Ocean sector has different behaviour again. The southern summer has a pattern strongly reminiscent of the theoretical description. However, in the northern summer the strong heating of the Eurasian continent dramatically accentuates the northward migration of the ITCZ. This leads to a radically different ocean–atmosphere interaction which we discuss in the next section.

2.10.4   *The Indian Ocean monsoonal circulation*

Monsoonal climates occur wherever there is a strong seasonal contrast between land and ocean temperatures. Their driving mechanism is the same as for a sea breeze, but on a gigantic scale. We will therefore start by examining why sea breezes occur.

During daylight hours the land surface is heated by the Sun. The warmed earth then heats the air above it, making it less dense than the surrounding atmosphere and causing it to rise. Near a coast where the sea surface temperature is cooler than the land, as is typical of summer, a pressure gradient will be established between the cooler, denser, air over the sea and the warmer, less dense, air over the land. Air will therefore flow in a layer from near the surface up to a few hundred metres in height, from the sea to the land. Aloft, the rising air over the land, and the removal of air from near the surface over the sea, creates a pressure gradient in the opposite direction. A circulation cell is thus created.

Sea breezes usually only affect a narrow coastal strip a few kilometres in width, but can extend inland for several scores of kilometres in particularly favourable conditions. Their small geographical extent means that the Coriolis force has only a small influence on the circulation. During summer evenings the circulation cell can reverse, as the land cools relative to the sea, giving a land breeze. Both sea and land breezes can only occur when the prevailing large scale, or synoptic, circulation is weak.

Monsoons are driven by similar forces. The Indian sub-continent and Tibetan Plateau are strongly heated during the northern summer, becoming substantially warmer than the air temperatures at equivalent altitude over the Indian Ocean[5]. This sets up a pressure gradient from the ocean towards India, but over much of the northern Indian Ocean. The net effect of this is shown in Fig. 1.6. The southern hemisphere Trade winds extend far north across the equator, the combination of the pressure gradient and the Coriolis force forcing the air to flow in a great arc across the Arabian Sea. This flow feeds into the giant sea breeze system, fuelling the rising air over southern Asia and providing the moisture for the torrential rainfall that this produces. The strongest flow is in a narrow jet at a height of 1–2 km which curves over East Africa. This has the properties of an atmospheric western boundary current, with the East African Highlands acting as the boundary. The land–sea contrast forces the summer northern movement of the ITCZ to be far greater than it would otherwise be. In the northern winter the equivalent 'land breeze' merely adds to the expected tropical circulation.

The Monsoon conditions force a number of special circulation features within the Indian Ocean itself. As the ITCZ crosses the equator during the equinoxal seasons the prevailing easterly winds weaken and reverse, giving westerlies. This produces a strong eastward-flowing jet at the surface of the ocean during these months, known as the Wyrtki jet after Klaus Wyrtki who led an international oceanographic expedition to the Indian Ocean in 1971 to investigate its circulation. This semi-annual variability in the

---

[5] This is particularly true of the Tibetan Plateau because of the much weaker attentuation of solar radiation at altitudes 4–5 km above sea level.

Fig. 2.35. The surface
ocean circulation of the
Indian Ocean in the
northern summer.
[Reprinted from G. L.
Pickard and W. J. Emery
*Descriptive physical
oceanography*, Copyright
(1982), page 209, with kind
permission from Elsevier
Science Ltd., The
Boulevard, Langford Lane,
Kidlington OX5 1GB,
UK.]

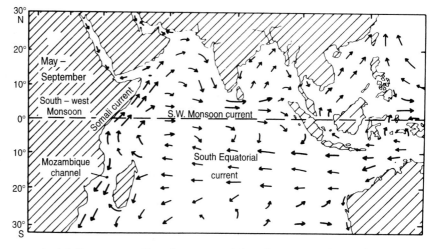

equatorial flow means that the equatorial under-current is weak and not always evident in this ocean.

In the Arabian Sea the sweep of the summer Monsoon winds parallelling the African and Arabian coasts leads to upwelling, particularly off Somalia and Oman. The winds induce an off-shore Ekman transport, as discussed in §2.10.2. A series of small circulation cells, or eddies, are also established along the African coast north of the equator, within the strong coastal Somali current. The surface circulation of the Indian Ocean in summer is shown in Fig. 2.35.

The Indian Ocean is not the only location where monsoon conditions occur. Northern Australia has a strong monsoon circulation which is part of the climatic cycle of El Niño and will be discussed in Chapter 5. Southeast Asia and China also experience monsoonal flows.

A summer monsoon occurs in west Africa, north of the equator, as well. The bulge of Africa that includes the Sahara, and extends out into the Atlantic ocean north of 10°N, is strongly heated during the summer. This draws the ITCZ unusually north in this region, bringing rains to the Sahel, which would otherwise be very arid. Disturbances to this pattern over the last 30 years have led to persistent drought affecting the Sahel. In §6.4.5 we will see that sea surface temperatures in the South Atlantic have decreased over this period, providing less energy and moisture to the southeasterly winds of the oceanic contribution to the ITCZ convergence.

2.10.5   *The polar regions*

One major oceanic impact on the atmosphere in the polar regions – reflection of solar radiation by pack ice – has already been discussed in various sections of Chapter 1. Where there is no pack ice, or frequent leads occur, the water is often warmer than the overlying air, providing a rich source of moisture, and hence latent heat, for the atmosphere. In sub-polar waters this frequently leads to the formation of active weather systems. The oceans are therefore major source regions for mid-latitude weather. This will be discussed in detail in §5.1.4.

The oceanic source of water vapour, coupled with the cold temperatures, leads to the polar oceans being particularly cloudy. Fog formation is also favoured. For example, sea smoke occurs where cold air is in contact with a warm sea which provides sufficient moisture to saturate the air. By contrast, in regions such as the Grand Banks off Newfoundland cold water flowing south from the Labrador Sea cools the air moving eastwards off a warming Canada in the spring to saturation. The abundance of latent heating potential in the atmosphere near the land–sea boundary in polar regions makes these regions particularly subject to the development of intense storms, of smaller scale than regular depressions, called polar lows.

The atmosphere, in turn, has a strong impact on the polar oceans. In §1.3.2 we saw how strong winter cooling and storm-force winds cause instability in the water column in the seas north of Iceland. The surface waters mix with much of the water column, creating deep water which spreads out to fill much of the bottom layers of the world's oceans. This is shown schematically in Fig. 1.14. Such deep water formation is sporadic and localised. Very limited observational evidence suggests that it occurs in small regions, or *chimneys*, only a few tens of kilometres in diameter. Similar processes occur in the Gulf of Lyon, in the NW Mediterranean, contributing to the characterisation of Mediterranean Water that eventually overflows into the North Atlantic. These have been better sampled.

The atmosphere also indirectly drives the formation of deep water originating from the Southern Ocean. This forms when sea water freezes to form sea-ice. The local release of salt, and consequent increase in surface water density, leads to relatively fresh but cold water sinking to the ocean floor to form Antarctic Bottom Water spreading as far north as 40°N in the Atlantic. Chapter 6 will investigate the surprisingly important links between the climate system and the origin and circulation of bottom water.

Formation of intermediate depth water is mostly driven by the atmosphere, as discussed in §2.10.1. Some regions of the polar oceans also contribute to intermediate waters by large-scale cooling, as in the case of water in the Labrador Sea, or cooling and Ekman convergence in the Southern Ocean (Antarctic Intermediate Water).

The atmosphere also drives the surface circulation. The most pronounced polar circulation is the global Circumpolar Current in the Southern Ocean, shown in Fig. 1.15. The driving force for this current is a continual, and strong, westerly wind stress. The Ekman transport is northwards, creating a slope downwards to the Antarctic coast in the isopycnals. The resulting off-shore pressure gradient is balanced by the Coriolis force to produce the strong, geostrophically controlled, Circumpolar Current. At the surface itself, the wind induces a northwards component to the flow, which leads to regions of strong latitudinal contrasts, or fronts.

The Circumpolar Current is strongly affected by the topography of the sea floor, and the intrusion into its zone of the landmass of South America. These boundary interactions produce regions of instability in the current, leading to copious formation of eddies which assist the mixing of water across the fronts.

2.10.6   *Oceanic eddies*

In several sub-sections of §2.10 we have encountered small circulation cells within the ocean called eddies. The meandering western boundary currents shed them during their unstable wandering. Seasonal eddies appear in the Somali current off eastern Africa. Eddies are created through instability in the Circumpolar Current. Throughout the world's oceans eddies can occur.

In the atmosphere the equivalent of eddies – the cyclones and anticyclones that make mid-latitude weather so variable and fascinating – are responsible for a major part of the poleward heat transport. In the ocean the eddies, while contributing to this climatic control, are very much the junior partners of the western boundary currents. The main reason for this ocean–atmosphere contrast is the relative size of the respective eddies. In the ocean eddies are scores of kilometres in diameter, in the atmosphere they can be more than a thousand. Atmospheric eddies also travel much faster, crossing the Atlantic in a few days for example, while a Gulf Stream ring may wander only a few hundred kilometres over a season.

Where eddies may be most important in redistributing heat through the oceans is off South Africa. The western boundary current flowing south along the East African coast – the Agulhas Current – overshoots south and west of South Africa before being swept into the influence of the Circumpolar Current. Much of its water then turns east, as shown in Fig. 1.15. However, some of the warm water transported from the Indian Ocean is spun off into eddies that enter the Atlantic. A numerical model of the Southern Ocean – the Fine Resolution Antarctic Model – developed by scientists from the United Kingdom, suggests that this occurs quite regularly every 4–5 months, with the Indian Ocean water being recognizable far into the Atlantic. A snapshot of the circulation and temperature field from this model at 120 m is shown in Fig. 2.36.

Arnold Gordon, of the Lamont–Doherty Geological Laboratory near New York, has proposed a Conveyor Belt model for global oceanic heat transport and water circulation, where water from the Pacific leaks through the Indonesian Archipelago, crosses the Indian Ocean and, via these eddies from the Agulhas Current, flows into the Atlantic. This transport is hypothesised to close the ocean circulation, the other half of the circuit being the deep water that flows slowly from the Atlantic to the Pacific. Gordon's model is illustrated in Fig. 1.14. In addition to providing a surface flux of water to the Atlantic, these eddies from the Agulhas Current also carry heat. The net heat transport of the Atlantic Ocean is estimated to be northwards at all latitudes, even south of the equator. The apparent anomaly in the South Atlantic (where heat would be expected to be transported polewards) could be explained by this Conveyor Belt theory. Consistent with this theory is the observation that heat flow in the Indian Ocean is everywhere southwards.

Eddies can also be biologically distinctive, compared to surrounding waters. Warm core eddies tend to be biologically poor, with a limited nutrient supply. Cold core eddies, by contrast, are very productive as there is a good supply of nutrients from the source region. In Chapters 3 and 4

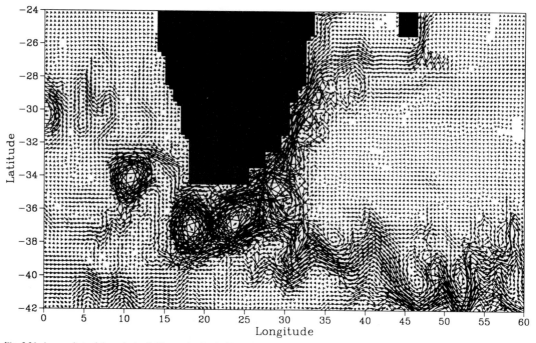

Fig. 2.36. A snapshot of the velocity field at a depth of 120m off South Africa from the Fine Resolution Antarctic Model. Note the strong Agulhas current flowing south along the east coast of Africa and the eddies of approximately 200km diameter spun off into the Atlantic. Picture courtesy of David Stevens.

we shall see that this may have implications for air–sea fluxes of various gases.

2.11   Oceanic impact on the marine atmospheric circulation

An intermittent filament of cloud bisects Fig. 2.32. The picture is a northern winter geostationary satellite photograph of the Atlantic Ocean. Being winter the filament – the ITCZ – passes south of the west African bulge. Over the land surfaces of both southern Africa and Brazil the ITCZ widens into an extensive cloud band. Swirling frontal systems can be seen crossing the North and South Atlantic. West of these fronts, and in the Trade wind belts, a multitude of small white patches show extensive cumulus growth in these moist, unstable, airstreams. Large areas of mid-latitude land are practically clear of clouds. These clear regions tend to extend over the adjacent seas to the west.

Many of these visible weather features are due to, or modified by, the oceans. The ocean provides the moisture and latent heat to generate storms, both of mid-latitude and tropical genre. It provides the moisture and warm surface to create the instabilities responsible for cumulus development. The upwelling on eastern sides of sub-tropical gyres is responsible for cooler sea surface temperatures. Resulting cooling of the overlying air makes the atmosphere denser and biases the position of the sub-tropical anticyclones to the eastern side of ocean basins, as can be seen in Fig. 1.7. This reinforces the upwelling further. The cooling effect of the sea surface can also lead to extensive fog banks, as is frequently observed in the Atlantic off the west coast of Namibia and Angola.

Conversely, some of the features of Fig. 2.32 are weaker over the sea. The ITCZ, though largely fuelled by moisture gathered over the oceans, is less organized and more patchy over the sea. Heating of the land surface provides an additional mechanism to accentuate the convergence of the Trade winds. The subsidence within the sub-tropical anticyclones is less general over the ocean because of the de-stabilising effect of a moisture supply.

We have already seen how the atmosphere and ocean can interact in monsoonal climates (§2.10.4). In Chapter 5 we will examine changes in the marine atmosphere during El Niño events, and the preferential formation of mid-latitude cyclones over the warm western sides of ocean basins. The ocean also influences the weather of adjacent land areas, giving them a milder, wetter climate than continental areas at the same latitude. This is particularly true on eastern coasts at mid-latitudes because the prevailing wind is blowing onshore. Maritime climates will also be considered further in Chapter 5.

There is, however, one atmospheric circulation feature which could not exist in the absence of the oceans. This is the hurricane.

2.11.1   *Hurricanes*

Hurricanes are intense atmospheric convection systems. They originate in the tropics, over warm seas, and are among the most devastating of natural phenomena. Every few years such a great storm roars out of the Bay of Bengal to kill perhaps hundreds of thousand of people in Bangladesh, as happened in 1970 and 1991. These deaths are mostly caused by drowning in the storm surges induced by the hurricane's winds, but massive damage, with resulting fatalities, occurs from the effects of these winds, from flooding by the torrential rainfall, and from the spread of disease caused by contaminated water and food supplies. Table 2.3 gives estimates of the fatalities in a number of historical hurricanes.

These storms can travel great distances, especially if their path does not cross over large areas of land. Atlantic hurricanes have been known to wreak havoc in the north-eastern United States. It is not unknown for the remnants of an intense hurricane to reach the United Kingdom, bringing heavy rain and strong winds, as shown by Hurricane Charlie over the August Bank Holiday weekend of 1986. In the western Pacific near Japan these storms are known as typhoons, while in the Phillipines they are called bagiuos. Near Australia, and in the Indian Ocean, such storms are known as tropical cyclones.

The maritime nature of hurricanes is shown by Fig. 2.37, a plot of the first reported location of such storms over a 20 year period. Also shown on this diagram is the summer hemisphere limit of the 26°C sea surface temperature isotherm. Hurricanes clearly form over very warm water between 5° and 25° from the equator. The energy supply is thus latent heat from

Table 2.3. *Fatalities in historical hurricanes*

Year	Death Toll	Location
1281	100 000	Kyushu, Japan
1737	300 000	Calcutta, India
1876	300 000	Chittagong, India
1881	300 000	Haiphong, India
1882	100 000	Bombay, India
1900	6000	Galveston, Texas
1942	11 000	Bengal, India
1963	6000	Cuba and Haiti
1970	300 000	Bangladesh
1977	10 000	Andra Pradesh, India
1991	140 000	Bangladesh

Fig. 2.37. Initial reported positions of all hurricanes during the 20 year period from 1961 to 1980. These are shown by dots. The solid line is the 26°C sea surface temperature in the summer hemisphere (i.e. July in the Northern Hemisphere and January in the Southern). Note that very few storms form in cooler waters, or less than 10° from the equator. The absence of hurricanes in the South Atlantic is due to the waters of the Brazil Current only exceeding 26°C by a small amount.

Hurricane distribution — 1961–1980

evaporation, and the Coriolis force is necessary to induce spin in a formative instability in the atmosphere. Hurricanes tend to form within the convergence regions on the eastern flank of travelling low pressure troughs in the Trade winds. The atmosphere needs to be convectively unstable, with little vertical shear in the wind. If such conditions are combined with a ready moisture supply from a warm ocean, and a divergent air flow aloft to accentuate the convection, a hurricane should develop.

Fig. 2.38 shows the main features of a mature hurricane. At the centre of the storm there is a small region without convection known as the eye. The wind speed here is much lower than in the main body of the hurricane,

Fig. 2.38. Model of a
mature tropical hurricane
(vertical cross-section),
showing principal cloud
formations and regions of
strong vertical motion. The
spiral rain bands are
indicated by hatching
under the clouds.

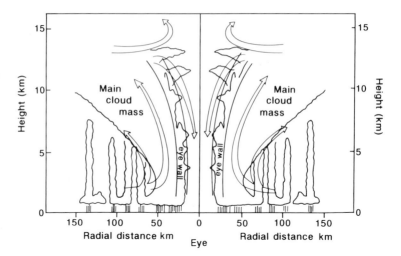

although not often the dead calm of some literary storms. There may also be
little cloud. The eye is some 20 km in diameter and has the lowest surface
pressure of the storm. This is typically well below 1000 mb, and has been
recorded as low as 870 mb in Typhoon Tip on October 12, 1979. Just
outside the eye is a narrow region of intense rainfall and wind speed in the
wall. This is the zone where wind speeds reach their highest values, possibly
more than 200 kmhr^{-1}. Away from the centre of the hurricane the wind
decreases, and the pressure rises, steadily. At radii of 100 km the wind speed
may still be over 60 kmhr^{-1} ($\sim$ 20 ms^{-1}). Another typical feature is the set
of spiral bands of strongly convective cloud. The existence of these was not
known until the use of radar for hurricane research at the end of World War
II. These spirals are regions of localised enhanced instability, producing
massive thunderstorms of cumulonimbus clouds, and intense rainfall.

A number of theories exist to explain the presence of the eye at the centre
of the storm. They can be conveniently combined as follows. As the
convection within a tropical storm intensifies, the air that is thus transported to the upper troposphere tends to spread out. This can be seen on a
small scale in individual mid-latitude thunderstorms from the presence of a
layer of thin and high ice cloud (cirrus) spreading in an anvil-shape ahead of
the storm. Of course, within a developing hurricane there is considerably
more air to be spread, and a positive pressure anomaly is produced at high
levels because of the ring of convection about the storm centre. This
accelerates the downflow, or subsidence, on the edge of the convection.
Subsidence warms the air by compression; try touching the valve on your
bicycle pump after you next inflate your tyres. This warming creates
buoyancy and acts to counter the subsidence. Eventually a state of
negligible vertical motion is achieved.

The spiral bands also show some interesting physics. The bands revolve
about the storm centre, but much slower than the wind speeds would
suggest. They are not, therefore, composed of the same air throughout their
existence. Parcels of air enter the unstable region, form thunderstorms, and

then leave. A typical passage time for the air is only 40 minutes. During its transit time the air parcel may lose water by precipitation at a rate of 30 mmhr^{-1}.

Hurricanes will decay when removed from their copious source of moisture. Thus, if one moves poleward over colder seas this supply lessens and the storm slowly decays. This is accentuated if it travels far enough to be diluted with mid-latitude air, although hurricanes can gain energy for a short time at this stage through interaction with the instabilities that cause extra-tropical cyclones. Even over tropical latitudes, if a hurricane moves over land it will lose energy rapidly. The central pressure can rise by several mb per hour. The moisture, and hence energy, supply is removed but the stored water is released, converting the hurricane into a very wet, if more conventional, depression.

Hurricanes feed from the warm tropical oceans. Their strong winds and rainfall will, in turn, have an impact on the ocean. The wind stress stirs the sea to deeper levels than normal. It has been suggested that the fisheries off the western Australian coast exist only because of the nutrients mixed into the surface waters by this occasional deep stirring. Such effects remain a moot point, however, as confirmatory observations are practically impossible in such extreme weather, and other work suggests that wind mixing alone may not have such penetrative power. This stirring will, however, lower the sea surface temperature, as will the rainfall, acting to dampen the storm development.

The high wind speeds also create local circulations in the ocean. This can be most important as the storm approaches land and pushes water in front of it. In shoaling seas this creates a storm surge, with water levels reaching as much as 5–10 m above normal sea level. These surges account for many of the fatalities from hurricanes, particularly in the Bay of Bengal where the seas are quite shallow near the Ganges delta. Storm surges are also produced by intense mid-latitude storms over confined basins, but the strong winds of hurricanes generally give higher floods. The worst surges in both cases are caused by a coincidence of the surge with high spring tides. The North Sea experiences minor surges in most winters from extra-tropical cyclones, with occasional disasters such as the 2 m surge in 1953 when 2000 people were killed in Holland and Great Britain. Surges are generally caused by strong alongshore winds, creating Ekman transport towards the coast, but interaction of this transport with the shoaling ocean floor can generate surges which move along the coast.

Further reading

A complete reference list is available at the end of the book but the following is a selection of the best books or articles to follow up particular topics within this chapter. Full details of each reference are to be found in the Bibliography.

Brown *et al.* (1989): An excellent introduction to basic physical oceanography.

Harvey (1985): A clear and simple discussion of basic oceanography and meteorology.

Jerlov (1976): A thorough introduction to ocean optics.

Pickard and Emery (1982): A good introduction to the descriptive aspects of physical oceanography.

Pond and Pickard (1983): Similar style to Pickard and Emery, but with a treatment of the dynamics of oceanography suitable for those with a little mathematics.

Simpson and Reihl (1981): A good introduction to hurricanes.

Tomzcak and Godfrey (1994): A very thorough survey of both the descriptive and dynamical aspects of oceanography, concentrating on regional variations.

3 Chemical interaction of the atmosphere and ocean

Both the ocean and the atmosphere are composed of a mixture of chemical compounds, some inert but many chemically very reactive. The two fluids therefore conceal an almost infinitely varied chemistry, which the presence of living organisms complicates even further. We are largely unaware of the molecular-scale world, where these processes occur, as we go about our every-day lives. Nonetheless, its invisible activity contributes fundamentally to the climate system, as we saw in Chapter 1.

The vast subjects of atmospheric and oceanic chemistry are well beyond the scope of this book. We will confine ourselves in this chapter to those processes occurring at the interface between the two media, and chemical contributions to the climate originating from an oceanic source. Some of these intimately involve biological activity; most interact with life forms in some way. To make the topic manageable those components strongly linked to marine biological processes will be left to the next chapter. There will, however, be a number of recurring themes: examples are cloud processes and the nitrogen and sulphur cycles.

A major part of the chemical interaction between the atmosphere and ocean begins with the transfer of chemical species, particularly gases, from one fluid to the other. This process depends on a number of physical and chemical parameters. The resulting control of the supply of oxygen and carbon dioxide to the ocean and atmosphere respectively lies at the root of much of the discussion of this, and the next, chapter. We will therefore begin by considering a key chemical property, namely, the solubility of gases in sea water.

3.1 Solubility of gases

The atmosphere is almost entirely a mixture of gases. These gases can enter the ocean across the air–sea interface. If no molecules of a particular gas were in solution in the upper ocean then this transfer would act as a drain on the atmospheric store of the gas in question (see §3.2). Once sufficient molecules from the atmosphere accumulate in the water for as many to leave the sea as enter in a given time, equilibrium is reached and the water is said to be saturated with respect to the gas. Atmospheric gases are generally close to a state of saturation in surface waters of the ocean. Indeed, their concentrations are often slightly above saturation because of the additional input to the ocean through the dissolution of bubbles from breaking waves.

All water in the sea has at one time been exposed to the atmosphere at the ocean surface. Therefore, it would be expected that atmospheric gases which are unreactive, such as nitrogen and the noble gases, would have similar concentrations throughout the world oceans. This has been found to be the case for all except a few gases. Helium has additional sources from hydrothermal activity and the radioactive decay of a product of nuclear bomb tests, *tritium*, ^3H. Tritium enters the ocean through its incorporation into water molecules in the atmosphere ($^3H^1HO$) and their subsequent precipitation or surface exchange. The oceanic concentration of helium therefore tells us more about the currents transporting water away from mid-ocean ridges, or diffusion into the interior of the ocean from the surface input of tritium, than its atmospheric content (§3.7).

Some atmospheric gases have been introduced by man. They can be entirely new additions to the environment, such as many of the *chloro-fluorocarbons*, or supplement natural sources, such as carbon dioxide. The distribution of their oceanic concentration helps to trace the movement of water from the surface into the deeper ocean.

The ocean will be a sink to those gases purely produced by man, such as the chlorofluorocarbons, or CFCs. Due to the chemical inertness of CFCs, these gases are emitted into the atmosphere at much greater rate than their slow absorption by the ocean. The size of the ocean sink for such gases is therefore rather small, and not considered further.

Some atmospheric gases, such as carbon dioxide, are very reactive in water and so the ocean can absorb rather more of these gases than their molecular weight and atmospheric composition would suggest. In Table 3.1 the ocean surface saturation concentrations for a selection of atmospheric gases at their present atmospheric partial pressures are given for two different temperatures. Table 3.1 also gives the solubility of these gases. This is defined as the saturation concentration that would be found if the entire atmosphere was composed of the particular gas. The solubility is an absolute measure with which to compare different gases, while the saturation value, which is just the product of the solubility and the atmospheric *partial pressure*, includes the effect of the actual atmospheric composition.

From Table 3.1 we can see several interesting properties of gas solubility. For inert gases it is a function of molecular weight. There is also, particularly for the heavier gases, a striking inverse dependence on temperature. This is because higher temperatures mean the molecules have a greater energy, and this, in turn, implies a higher rate of exchange of molecules between the air and water, so that the equilibrium situation occurs with fewer molecules of gas within the liquid. This dependence on temperature is illustrated for a number of gases in Fig. 3.1.

Some atmospheric gases have anomalously high solubilities, for their molecular weight. These include carbon dioxide, methane, and nitrous oxide; all of these are greenhouse gases. Their high solubility occurs because the gases are involved in reactions with water which convert much of the gas to soluble ions. Such gases also show an enhanced temperature-

Table 3.1. *Saturation concentration and solubility for atmospheric gases at 1 atmosphere pressure in sea water of salinity 35*

Gas	Weight g/mole	Partial pressure (air) atmos.	Oceanic saturation (ml/l*) 0°C	24°C	Solubility (ml/l) 0°C	24°C
Nitrogen (N$_2$)	28	0.781	14.3	9.2	18.3	11.8
Oxygen (O$_2$)	32	0.209	8.1	5.0	38.7	23.7
Argon (Ar)	40	9.3×10^{-3}	0.39	0.24	42.1	26.0
Carbon Dioxide (CO$_2$)	44	3.54×10^{-4}	0.51	0.24	1437	666
Neon (Ne)	20	1.8×10^{-5}	1.8×10^{-4}	1.5×10^{-4}	10.1	8.6
Helium (He)	4	5.2×10^{-6}	4.1×10^{-5}	3.8×10^{-5}	7.8	7.4
Krypton (Kr)	84	1.1×10^{-6}	9.4×10^{-5}	5.1×10^{-5}	85.6	46.2
Nitrous Oxide (N$_2$O)	44	3.0×10^{-7}	3.2×10^{-4}	1.4×10^{-4}	1071	476

*A more consistent unit would be mol/kg, but as the values are several orders of magnitude smaller for these units I use the more traditional value for ease of comparison.

dependence of their solubility (compare the variation of solubility of CO$_2$ and O$_2$ with temperature in Fig. 3.1). This enhancement occurs because the rates of the chemical reactions that are responsible for allowing more gas to be absorbed are also temperature-dependent. If these chemical reactions can soak up gas faster than it can enter the ocean then saturation will not be reached. It can also be found that in particular locations excessive super-saturation occurs due to chemical or physical input of gas. Fig. 3.2 shows the partial pressure of CO$_2$ in the surface waters of the ocean. In §3.3 we will discuss in detail the uptake of carbon dioxide but note for now the excess near the equator and the depletion in sub-polar and higher latitudes.

Some gases which are present in only very small quantities in the atmosphere are products of biological activity within the sea. For these the atmosphere is a sink, rather than a source, region. Such gases, which can interact with the climate through contributing to the greenhouse effect (for example, N$_2$O and CH$_4$) or cloud processes (for example, dimethyl sulphide), will be discussed in Chapter 4.

Solubility is also weakly dependent on salinity, with higher salinity decreasing solubility. Pressure is also important. Deeper in the ocean, where the pressure is higher, a higher concentration of gas can be sustained. We will see how this is important for air–sea exchanges in §3.3.2.

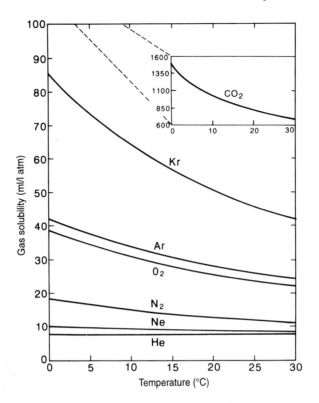

Fig. 3.1. The solubility of various atmospheric gases in sea water as a function of temperature. The temperature range covers most of the observed range in oceanic waters. Units are millilitres of gas contained in a litre of sea water of salinity 35 psu, assuming an overlying atmosphere purely of each gas. Data from Broecker and Peng (1982).

3.2   Gas exchange across the air–sea interface

The solubility of a gas determines the relative ease with which it can be absorbed into the ocean if there were no other gases present in the atmosphere, and if an equilibrium state was achieved. The troposphere, however, is a mixture of many gases and the basic driving force for exchange of gas across the air–sea interface is the difference in gas concentration, or partial pressure, between the two media. The flux, F, of a gas across the interface into the ocean is often written as

$$F = k_T(P_a - P_s) \tag{3.1}$$

where P_a and P_s are the partial pressures of the gas in question in the atmosphere and ocean respectively and k_T is the *transfer velocity*.

As its name suggests k_T has the units of a velocity; it represents the variability of the rate of exchange due to the sea state and atmospheric stability. A calm sea and stable air will allow only slow exchange because the surface air mass is renewed infrequently and there is little bubble entrainment in these conditions (see §2.8.2). By contrast, rough seas and strong winds allow frequent renewal of the surface air from above and also actively bypass molecular diffusion across the interface by copious bubble production. Intermediate conditions show moderate transfer. There is often an abrupt change in sea state, and transfer velocity, when the wind becomes strong enough for breaking waves. The wind speed is thus taken as an

Fig. 3.2. Distribution of the partial pressure of CO_2, in microatmospheres, in the surface waters of the global ocean, relative to the atmospheric partial pressure at the time of measurement. Negative values indicate sub-saturation (i.e. the surface water has a lower partial pressure than the atmosphere); positive values show super-saturation. Dotted lines indicate estimated rather than observed values. [Adapted, with permission, from C. D. Keeling, *J. Geophys. Res.*, 73, 4543–53 (1968), copyright of the American Geophysical Union.]

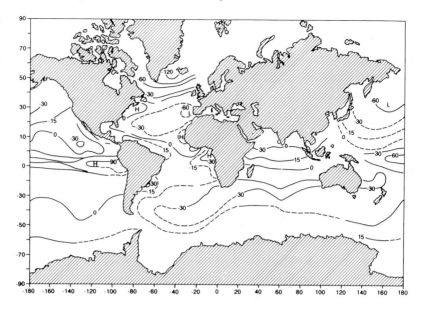

indicator of these physical states and the variation in k_T, and thus gas exchange rate, is a strong function of this variable, as shown in Fig. 3.3.

There are also other effects which the transfer velocity must take into account. The large solubility differences between different gases, due partially to molecular weight but mostly to a gas's chemical reactivity in water, have important consequences for the exchange rate at low wind speeds. At high wind speeds the supply of gas is regulated by the physical mechanisms associated with breaking waves rather than chemical uptake. Although not clear from Fig. 3.3, at wind speeds below 4–5 ms^{-1} chemically reactive carbon dioxide has a 50% higher transfer velocity than oxygen. However, in the breaking wave regime inert gases are pumped into the sea at a greater rate than chemically active ones, because the air immediately above the water surface, which is entrained during the wave breaking, is enriched in the inert species.

Another mechanism which affects the air–sea flux rate is the heat transfer between the two media. If the air is humid and the latent heat flux is directed towards the ocean, producing condensation on the water surface, then gas flux into the ocean is inhibited. It has recently been shown that the reverse is true for sensible heat. Essentially, the solubility increase with decreasing temperature establishes a chemical potential gradient towards the ocean, if the sea surface temperature is less than the air temperature. The resulting flux is then

$$F = -\frac{k_T}{RT_m}P_m Q_{sol}\left(\ln\left(\frac{T_s}{T_a}\right) + \ln\left(\frac{P_s}{P_a}\right)\right) \tag{3.2}$$

where P_a is atmospheric partial pressure, P_s the ocean surface partial

Fig. 3.3. Variation of the
gas transfer velocity with
wind speed, *u*. Note the
difference in wind
dependence between the
two gases O₂ and CO₂.
The transfer of oxygen
effectively obeys the
breaking wave regime at a
wind speed some 2 ms⁻¹
less than for carbon
dioxide. The units of
transfer velocity are
equivalent to the cm of air
column entering the water
per hour.

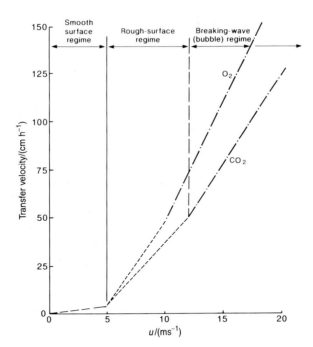

Fig. 3.3. Variation of the gas transfer velocity with wind speed, u. Note the difference in wind dependence between the two gases O_2 and CO_2. The transfer of oxygen effectively obeys the breaking wave regime at a wind speed some 2 ms^{-1} less than for carbon dioxide. The units of transfer velocity are equivalent to the cm of air column entering the water per hour.

pressure, $P_m = 0.5(P_a + P_s)$, $T_m = 0.5(T_a + T_s)$, R is the ideal gas constant, and Q_{sol} is the energy released when a mole of the gas is dissolved in sea water, termed the *enthalpy of solution*. Equation (3.2) reduces to (3.1) if the air and sea temperature are identical. Much of the ocean equatorward of 45°, for much of the year, is warmer than the air above it. The chemical gradient decreases the flux of gas into the sea in these circumstances by about 10% on average. However, over polar latitudes, particularly in summer, over regions of strong upwelling, and off the east coast of heated continents during summer the sensible heat gradient acts to enhance gas flux into the ocean.

The transfer velocity is often termed the piston velocity because it can be thought of as the size of a column of gas pumped into, or out of, the ocean by the gas partial pressure difference across the air–sea interface. This is an appealing physical association for k_T. There is a danger, however, of this association gaining too strong a hold on the scientific imagination. The preceding discussion clearly shows that the transfer velocity is dependent on a large number of physical parameters. The applicability of present gas exchange theory is thus hedged with a multitude of special cases. The transfer of gases, particularly carbon dioxide, across the air–sea boundary is an important climatic process. Accurate estimates of the spatial and temporal fluxes of gases are thus vital for well-based predictions of the future climate. Table 3.2 shows present estimates of the fluxes of climatically important gases between the ocean and atmosphere. Water vapour has been neglected because of the essential balance between precipitation and evaporation over the global ocean. As a comparison, the evaporation of water vapour from the ocean each year is 4.25×10^5 gigatonnes.

Forecasts of future change assume partitioning of carbon dioxide

Table 3.2. *Annual, global fluxes of gases between the ocean and atmosphere. A positive value is into the ocean*

Gas	Flux (gigatonnes/yr)
Carbon Dioxide	2.0 ± 0.8
Methane	-0.010 ± 0.005
Nitrous Oxide	-0.0020 ± 0.0006
Sulphur Gases (e.g. DMS)	-0.030 ± 0.010
Non-methane hydrocarbons	-0.065 ± 0.030

between the air and sea will remain consistent with current behaviour. This may be a source of considerable uncertainty, because of our presently poor knowledge of the oceanic control of this greenhouse gas (see §3.3.2). While estimates of the transfer velocity are available in a number of controlled situations the complexity of the processes involved are such that global estimates of the flux of CO_2 into the oceans are uncertain by a factor of 50% at least. The physical and chemical processes involved in such fluxes need to be better understood in combination. This would ultimately lead to a more useful theory of gas exchange, bringing the quantum step in accuracy of our flux estimates required for confidence in predictions of climatic change.

3.3 The carbon cycle

Table 3.2 showed that the ocean is a net sink for carbon dioxide, while it is a net source for most of the other greenhouse gases. In this section we shall explore the reason for this behaviour, but first we will examine the role of this oceanic sink in the full carbon cycle.

3.3.1 The carbon cycle

Carbon cycles between the Earth, biosphere, ocean and atmosphere, as shown schematically in Fig. 3.4. There is considerable uncertainty over the magnitude of the storage that is currently occurring in several of these reservoirs. Carbon dioxide is mixed through the troposphere within a few months. Our understanding of this component of the carbon cycle is therefore fairly well established. However, the uncertainty in the other storage components is of a similar size to their estimated store.

The variation in atmospheric CO_2 is shown in Fig. 3.5, a monthly time series taken at Mauna Loa in Hawaii. A rapid seasonal cycle is superimposed on a longer term more gradual increase. The latter trend is due to anthropogenic activity. The annual modulation is due to the biosphere. During the summer large amounts of carbon are fixed in the terrestrial and oceanic biosphere by photosynthesis (see §4.1). The limited photosynthesis of winter leads to net respiration and release of CO_2 to the atmosphere. Over the globe the seasonal exchange of CO_2 between the oceans and atmosphere is roughly in balance between the two hemispheres. However, the greater land mass of the northern hemisphere, as compared to the southern, leads to a greater terrestrial uptake of carbon during the northern

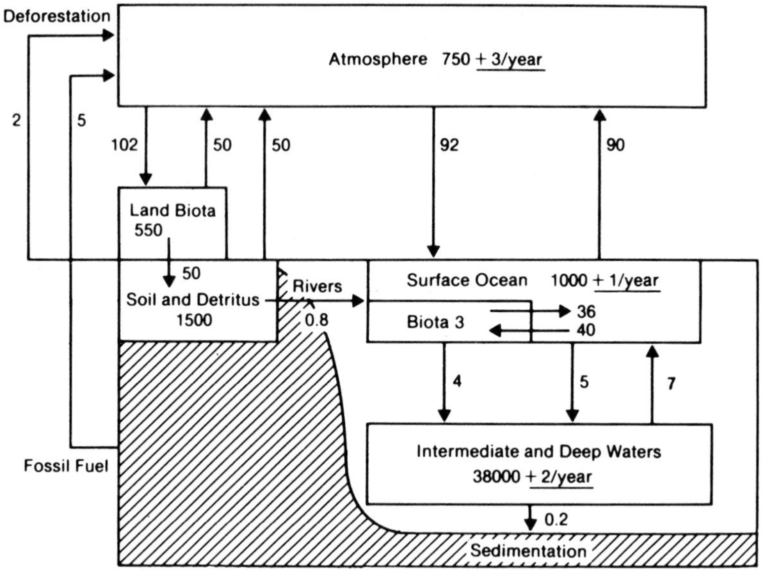

Fig. 3.4. Global carbon reservoirs and annual fluxes. Numbers underlined represent net annual CO_2 accumulation due to human action. Units are gigatons (10^9 metric tons, or 10^{12} kg) of carbon in the reservoirs and GtCyr^{-1} for fluxes. [Fig. 1.1 of Houghton *et al.*, (1990). Reproduced, with the permission of Cambridge University Press.]

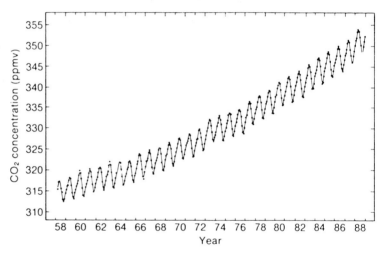

Fig. 3.5. Monthly average atmospheric CO_2 concentrations, observed continuously at Mauna Loa, Hawaii. [Fig. 1.4 of Houghton *et al.*, (1990). Reproduced, with the permission of Cambridge University Press.]

summer than is released over land in the winter of the southern hemisphere. An annual cycle in global atmospheric CO_2 concentration results, with a peak during the northern winter's respiration phase.

The carbon cycle, and the changes in distribution of carbon between the different reservoirs, is an important component of the climate system. Much carbon appears to be sequested into very long-term stores, in the interior of the Earth, sediments and deep oceans. Chapter 6 will show, however, that large changes in atmospheric, and therefore climatically active, carbon dioxide can occur naturally. A near-doubling of CO_2 occurred from 20 000 years BP, during the height of the last glacial period, to 1750 AD (i.e. 200 BP), before man's influence became noticeable. Substantial natural redistribution of CO_2 between the atmosphere, ocean and biosphere is therefore possible. Chapter 4 will consider some of the biosphere's contribution to this exchange; Chapters 6 and 7 will discuss natural and anthropogenic changes to the carbon cycle in more detail. A very important component of this exchange, however, is the storage mechanism in the ocean. This is largely of chemical origin and discussed in detail in the next section.

3.3.2 Oceanic control of carbon dioxide – principal processes

Carbon dioxide is a major product of the combustion of carbon-containing substances, such as wood, coal and oil. Since the dramatic increase in combustion initiated by the Industrial Revolution in the late eighteenth century the atmospheric concentration of CO_2 has increased by about a third (see Fig. 7.6). Substantially less carbon dioxide has remained in the atmosphere than has been emitted into it since 1750, perhaps as little as 50%. Much of this extra carbon dioxide is thought to have been absorbed by the ocean, as a result of the gas's high solubility. The principal reason for the ocean being a sink of carbon dioxide is shown in (3.3):

$$CO_2(gas) + H_2O \ (\rightleftharpoons H_2CO_3) \rightleftharpoons H^+ + HCO_3^- \rightleftharpoons 2H^+ + CO_3^{2-} \quad (3.3)$$

The component reactions of (3.3) are fast, and the system exists in equilibrium. Thus, additional CO_2 will be quickly swept into this system, distributing carbon atoms between the gas phase and the carbonate (CO_3^{2-}) and bicarbonate (HCO_3^-) ions in solution. The reason for the rapidity of reaction (3.3) can be seen by examining the reaction rates, or *dissociation constants*, K_1 and K_2 of two equations of the full equilibrium in (3.3), namely

$$K_1 = \frac{[H^+][HCO_3^-]}{[H_2O][CO_2]} \quad (3.4)$$

for the gas to bicarbonate equilibrium in the left half of (3.3), and

$$K_2 = \frac{[H^+][CO_3^{2-}]}{[HCO_3^-]} \quad (3.5)$$

for the bicarbonate to carbonate equilibration to the right of (3.3). In (3.4) and (3.5) the square brackets represent the concentration of the respective chemical species. The dissociation constants depend on temperature, pressure and salinity. Table 3.3 shows values for K_1 and K_2 at different temperatures, for a pressure of 1 atmosphere and a salinity of 35 psu. The greater magnitude of K_1 means that most CO_2 is converted to HCO_3^- and only a very small portion, is then converted to the carbonate ion as K_2 is roughly a thousand times smaller than K_1. The net effect of the reactions in (3.3) therefore leads to the summary reaction

$$CO_2(gas) + H_2O + CO_3^{2-} \rightleftharpoons 2HCO_3^{2-} \quad (3.6)$$

The bicarbonate/carbonate species are not produced solely from the equilibrium with carbon dioxide, but also have a source from deposition, by rivers or wind-blown dust, of the weathering products of rocks containing calcium carbonate ($CaCO_3$). Limestone is one common source of such

Table 3.3. *Dissociation constants for the principal reactions in (3.3), at a pressure of 1 atmosphere and a salinity of 35 (after Broecker and Peng, 1982).* K_1 *is multiplied by the constant* $[H_2O]$

Temperature (°C)	K_1	K_2
0	6.73×10^{-7}	3.53×10^{-10}
5	7.17×10^{-7}	4.04×10^{-10}
10	7.98×10^{-7}	4.72×10^{-10}
15	8.73×10^{-7}	5.58×10^{-10}
20	9.41×10^{-7}	6.59×10^{-10}
25	10.00×10^{-7}	7.69×10^{-10}
30	10.47×10^{-7}	8.77×10^{-10}

material. This 'pre-existing' oceanic carbonate source from calcium carbonate weathering permits a greater absorption of carbon dioxide than would otherwise occur. This can be seen from considering the dissociation constant summarising the entire process in (3.6):

$$K = \frac{[HCO_3^{-}]}{[CO_2][H_2O][CO_3^{2-}]} = K_1/K_2 \tag{3.7}$$

For a fixed temperature and pressure K is a constant, so weathering-enhanced concentrations of carbonate, and hydrogen ions, through the two right-hand equilibria of (3.1), must draw more carbon dioxide into solution, as $[H_2O]$ is too large to be affected by the reactions of (3.3).

In surface waters, biological processes involving the interaction of phosphate, nitrate and carbon lead to acidity and total dissolved carbon amounts being approximately constant over the globe (these processes will be pursued in Chapter 4). The numerator of (3.7) will therefore be constant, as is $[H_2O]$. However, the reaction rate, K, will depend on the temperature, becoming slower with warmer temperatures. If the atmospheric carbon dioxide is in equilibrium with the surface water, then

$$[CO_2] = S_{CO2}\, p_{CO2} \tag{3.8}$$

where S_{CO2} is the solubility of carbon dioxide and p_{CO2} is the partial pressure of carbon dioxide in the over-lying atmosphere. This means that the partial pressure can be expressed by

$$P_{CO2} = \frac{1}{KS_{CO2}} \frac{[H^{+}]^2[CO_3^{2-}]}{[H_2O]} \tag{3.9}$$

Equation (3.9) suggests that, in equilibrium conditions, the atmospheric partial pressure of carbon dioxide should be a strong function of temperature. The relative changes of the solubility and K with temperature imply that the partial pressure should decrease by about a factor of three between 24°C and 0°C. We have, however, already seen that the atmospheric carbon dioxide concentration is essentially uniform over the globe. The atmosphere mixes fast enough for this thermally-driven poleward gradient not to be established.

There cannot, therefore, be a general equilibrium of the carbon dioxide exchange between the ocean and atmosphere. Fig. 3.2 shows that the CO_2 levels in the polar surface waters are lower, while the tropical water CO_2 partial pressure is higher, than the equilibrium value for the over-lying atmosphere. There must be a net flux of atmospheric carbon dioxide into the polar waters, and out of the tropical oceans. This flux does not push the ocean towards equilibrium with the atmosphere, because the temperature of the water is changing due to the march of the seasons and changes in the surface ocean currents, and also because there are exchanges of the surface water with deeper waters in both regions, as discussed in Chapters 1 and 2. There is a competition between the equilibration time for carbon dioxide and the time scale for surface water stagnation which ensures that equilbrium conditions are not always achieved.

The reaction sequence in (3.3) is the principal way in which carbon dioxide intake into the ocean is chemically enhanced. However, there are other reactions which might be significant. These are reactions of CO_2 with marine minerals, either in bottom sediments or in solution within the water column. The breakdown of calcium carbonate fragments, which can include the shells of crustaceans, by the reaction

$$CaCO_3 + CO_2 + H_2O \rightleftharpoons Ca^{2+} + 2HCO_3^- \qquad (3.10)$$

is one such process. This is the same reaction as is involved in the weathering of limestone on land. It is not thought to be widely important as a carbon dioxide sink because surface waters are generally super-saturated with respect to solid phase $CaCO_3$. This inhibits (3.10) from occurring. However, in bottom waters in coastal zones, where anthropogenic carbon dioxide is easily able to penetrate, the water appears to be under-saturated with respect to impure forms of calcium carbonate. It is estimated that perhaps 1.5% of the carbon dioxide produced by man each year could be absorbed by the oceans in coastal regions by reaction (3.10).

We saw in §1.5 that the oceans are net absorbers of the additional carbon dioxide added to the atmosphere each year by human activity, although considerable uncertainty continues over the apportionment of the carbon dioxide that leaves the atmosphere for the ocean or land. Fig. 3.4 shows that much of the CO_2 that enters the ocean from the atmosphere does not remain in the surface waters but is transferred by mixing, and formation of deep water, into the intermediate and deep ocean. Much of the additional anthropogenic carbon dioxide will also follow this route, forcing the surface ocean to move towards equilibration with the increasing carbon dioxide of the atmosphere at an even slower rate. The long time scales of deep water overturning – several centuries to a thousand years – coupled with the greater solubility of carbon dioxide in water under pressure, means that the ocean acts as a giant reservoir of carbon dioxide, and a drag on climatic change associated with CO_2 increase. The converse of this also exists. If atmospheric CO_2 levels were suddenly decreased, then the ocean would slowly leak carbon dioxide back to the atmosphere, acting to push the climate back towards its previous state.

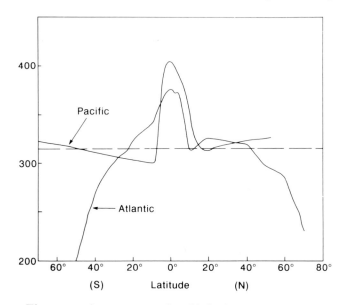

Fig. 3.6. Cross-section of surface p_{CO_2} in the Atlantic and Pacific Oceans, observed on GEOSECS cruises of the 1970s. The units are microatmospheres, or parts per million, of CO_2. The dashed line shows the atmospheric concentration at that time.

The ocean also exerts another, biological, control on atmospheric carbon dioxide. As we shall see in Chapter 4, input of CO_2 to the ocean encourages biological activity. Some of this carbon is then stored within marine life-forms, where its return to the atmosphere is delayed. If the marine organisms add to the sediment on the ocean floor upon their death, then the carbon is added to the geosphere's reservoir. This may not be re-cycled to the atmosphere for millions of years.

3.3.3 Oceanic control of carbon dioxide – geographical variations

The poleward decrease of upper ocean p_{CO_2} is not without caveats (Fig. 3.2). In Fig. 3.6 latitudinal sections of ocean surface p_{CO_2} obtained in the early 1970s for the Atlantic and Pacific Oceans show both similarities and differences. Both oceans show distinct peaks near the equator, with the Atlantic peak being rather broader. Both also show the ocean and atmosphere in near-equilibrium into the mid-latitudes. However, their behaviour at high latitudes is distinctly different. The Pacific p_{CO_2} remains near equilibrium all the way to Siberia and the Antarctic. In the sub-polar Atlantic, however, there is a distinct decrease in both hemispheres.

The poleward difference between the two great ocean basins is a direct result of the oceanic circulation. In the far North Atlantic, and the Weddell Sea off Antarctica, bottom and intermediate waters are formed. This takes surface water into the deep ocean, and continually renews the water exposed to the surface. This cold, carbon-dioxide-poor water then takes up CO_2 from the atmosphere, but is mixed down before reaching equilibrium. The Pacific circulation does not have regions of deep water formation, even in the Southern Ocean[1]. There is also less formation of intermediate water

---

[1] Occasional deep water formation may occur in the Ross Sea in the Antarctic sector of the South Pacific and north of the Aleutian Islands in the North Pacific.

Fig. 3.7. Potential CO_2 pressure as a function of depth in the equatorial Pacific. Potential CO_2 pressure is the CO_2 partial pressure a sample of water would achieve if it were depressurized to 1 atmosphere pressure and warmed to 25°C, without changing the chemical composition of the sample. [Fig. 3.25 of Broecker and Peng (1982). Reproduced with permission of W. S. Broecker.]

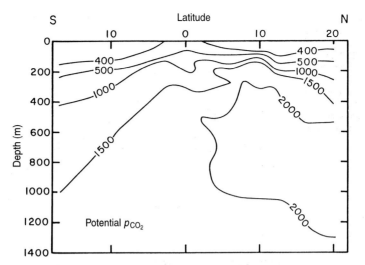

in the North Pacific gyre. The oceanic carbon dioxide flux is therefore closer to equilibrium in the slower vertical mixing regime of the sub-polar Pacific. The sub-polar Atlantic Ocean is thus the major oceanic sink of atmospheric carbon dioxide.

The 20% increase in surface oceanic carbon dioxide content at the equator is also due to the circulation. In §2.10.3 we saw that Ekman divergence in the upper ocean caused upwelling along the equator. This upwelled water is colder than characteristic tropical sea surface temperatures, giving it a higher solubility. It has also come from regions of higher pressure; this means that the source water for the upwelling has a compressed, and so higher, quantity of carbon dioxide. The effect of decreasing the pressure as the water rises to the surface is to let the CO_2 expand. The water itself, however, is only slightly compressible, compared to a gas, and so it changes its volume little. The net effect is to increase the upwelling water's carbon dioxide concentration. Fig. 3.7 shows the potential p_{CO_2} if water of different depths and locations in the eastern Pacific were raised to the surface. Upwelling of water from as close to the surface as 100 m would be sufficient to give p_{CO_2} values well above those observed. These are not at their potential level because carbon dioxide is given up to the atmosphere rapidly in these regions, and much carbon is gathered from the mixed layer by plankton. The upwelling occurs too quickly for equilibrium with the atmosphere to be achievable, and for a significant anomaly in p_{CO_2} to be maintained. Upwelling rates of several tens of metres per year have been estimated to be required to maintain the observed imbalance.

The relative stability of the tropical atmospheric, and hence surface oceanic, circulation in the Pacific leads to the large but equatorially-confined p_{CO_2} anomaly. By contrast, in the Atlantic there is considerable latitudinal shifting of the ITCZ through the year. This causes the dynamics of upwelling to vary seasonally in strength, and leads to a less dramatic, but more extensive, anomaly.

While the mean latitudinal variation of p_{CO_2} is as shown in Fig. 3.6 there are further local variations of climatic significance. Upwelling regions in

Fig. 3.8. Geographical distribution of anomalous total CO_2 in the surface waters of the Atlantic, based on measurements of salinity, phosphate, nitrate, and alkalinity during GEOSECS and TTO expeditions. The total CO_2 has been corrected for fossil fuel emissions. Units are μmol/kg. Data from Broecker and Peng (1992).

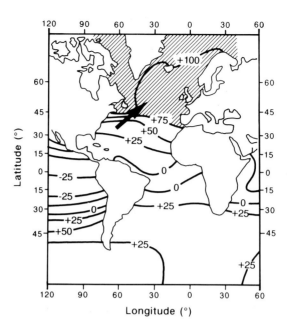

general will display properties similar to the equatorial zone, and thus have high carbon dioxide contents. This is evident in Fig. 3.2 in the east Atlantic and northwestern Indian Ocean, for instance.

The warm western boundary currents are also strikingly anomalous regions. The warm water, while at equilibrium with the atmosphere when it begins its poleward travel, is poor in dissolved carbon dioxide, because of the low solubility at such temperatures. As it moves poleward the water is cooled, allowing more carbon dioxide to enter solution. The rapid poleward flow tends to lead to faster cooling than the equilibration processes can keep up with. There should therefore be considerable CO_2 uptake in areas of the oceans such as the Gulf Stream and the Kuroshio Current. This is consistent with Fig. 3.8. a map of the upper ocean total carbon content for the Atlantic. Total carbon is defined as the sum of dissolved carbon dioxide, bicarbonate and carbonate ions.

3.4   Oxygen in the ocean

Oxygen is in a state of super-saturation in surface waters, mostly because of entrainment of bubbles from breaking waves (see §2.8.2). As these bubbles of air are carried below the surface, the increase in pressure forces gas into solution. Hence not only oxygen but most atmospheric gases are slightly super-saturated at the ocean surface. There is also biological production of oxygen in parts of the water column where there is sufficient light for photosynthesis (see Chapter 4). This is almost balanced by oxygen consumed by respiration. As light levels decrease with depth through the ocean the balance swings further towards consumption of oxygen rather than production, and eventually there is only respiration. Oxygen therefore

Fig. 3.9. Variation of oxygen concentration with depth at four locations during the GEOSECS cruises of 1972–3. Site (a) is in the Norwegian-Greenland Sea (74° 55′N, 1° 07′W); (b) is in the western sub-tropical Atlantic (31° 48′N, 50° 46′W); (c) is in the tropical east Atlantic (10° 59′N, 20° 32′W); (d) is in the central North Pacific (31° 22′N, 150° 02′W). The units are micromoles of oxygen per kilogram of sea water.

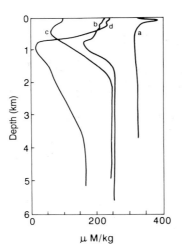

Fig. 3.10. Difference between the observed and saturation O_2 contents for surface waters in the equatorial zone in the central Pacific. Two transects are shown, both of which reveal the usual super-saturation of surface waters in oxygen and the equatorial deficit due to upwelling of oxygen-poor water. [Fig. 3.7 of Broecker and Peng (1982). Reproduced with permission of W. S. Broecker.]

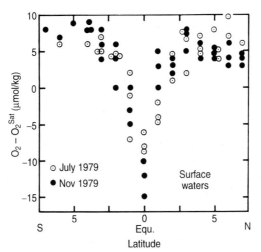

tends to reach a minimum at some depth below the surface. This depth dependence of the dissolved oxygen level is illustrated in Fig. 3.9.

This is only a summary of what occurs. In Chapter 4 we will investigate the biological processes in more detail, particularly the gaseous by-products of life which influence the climate through precipitation mechanisms and the greenhouse effect. In the remainder of this section we will examine some geographical anomalies to the simple pattern of oxygen concentration described in the last paragraph.

Fig. 3.10 shows a section of surface oxygen levels, compared to saturation, northward through the tropical east Pacific. These levels are close to 5% saturation, for the reasons discussed above, but within 100 km of the equator the oxygen level drops dramatically, becoming unsaturated. From our discussion in §2.10.3 of the tropical ocean circulation, and the enhanced equatorial levels of surface p_{CO2} discussed in the last section, you can probably deduce the cause for this equatorial anomaly. The upwelling brings water from depth, which is not only rich in carbon dioxide, but depleted in oxygen through previous biological activity. Oxygen levels might therefore be expected to be depressed in most upwelling regions. The under-saturation in oxygen is not sufficient, however, to decrease the biological activity of these regions. Indeed, upwelling regions are usually very productive, because of the transport of nitrates and phosphates – the food supply for plankton – to the surface. This increased productivity leads to pronounced oxygen minima at depth beneath upwelling regions. In a few areas of the global ocean – the Arabian Sea and the tropical eastern margins of the Pacific – almost all the oxygen is consumed at the thermocline. In §4.2.2 we will see that this has implications for the ultimate fate of a greenhouse gas, nitrous oxide, produced by biological activity.

Fig. 3.9 showed oxygen levels rising below the thermocline, as well as the general decline with depth in the upper ocean. This reversal occurs because of the transport of oxygen-rich water from the polar regions of the North and South Atlantic. When this deep water is formed it will be saturated with oxygen. Moreover, as this water is cold it will have oxygen levels about 60% higher than those in the tropics; Table 3.1 and Fig. 3.1 illustrate this temperature dependence of solubility. Some of this oxygen is naturally consumed by biological processes in the formation regions, but only 5–10%. Less is consumed in the North Atlantic because the time scale of deep water formation is rather faster there. As the deep water spreads to fill the world ocean slight decreases in this oxygen level occur, due to the respiration of the small amount of life in the deep ocean. This decrease is a major piece of evidence for the deep ocean circulation hypothesised in §1.3.2 and Fig. 1.14. There is a clear decrease in oxygen levels between the Atlantic and the Indian and Pacific Oceans, shown in Fig. 3.11. This deep circulation provides the longest time scales, of centuries, for the ocean's influence on climate.

3.5   The transfer of particles

The ocean surface is subject to a continual barrage of particles from the atmosphere. Some of these owe their origin to previous expulsion from the ocean surface, as discussed in §2.8.2. Others derive from chemical and physical processes within the atmosphere causing coagulation of smaller particles or gas molecules. Yet more have been swept, or thrown, up from a land surface and carried perhaps thousands of kilometres in the atmospheric circulation.

The cascade of these particles into the sea affects the chemistry of the ocean in many ways. For some of the trace constituents this will be their major source. However, in this chapter we are not studying the general chemistry of the ocean, but only those aspects relevant to climate. In this context our interest will be restricted to those particles likely to affect the nutrient supply of phytoplankton or the cycling of greenhouse gases.

Particles are also being continually ejected from the ocean by evaporation, bubble bursting and spray. We have already considered the physical processes involved in this production of marine aerosol (§2.8.2); here we will

Fig. 3.11. Distribution of Actual Oxygen Utilization (AOU) in the deep ocean, at a depth of 4000m. AOU is the saturation oxygen content minus the measured oxygen content. Higher values indicate greater use of oxygen by biological organisms. Two sources of deep water are seen, from the North Atlantic and Weddell Seas. [Fig. 3.8 of Broecker and Peng (1982). Reproduced with permission of W. S. Broecker.]

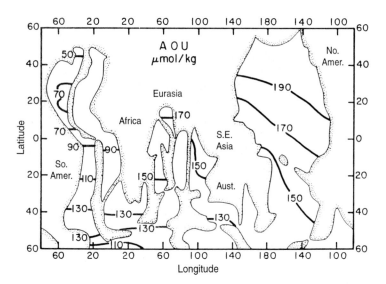

Fig. 3.12. Aerosol size-distribution spectra characteristic of continental and maritime air. Using data from Ludlam (1980).

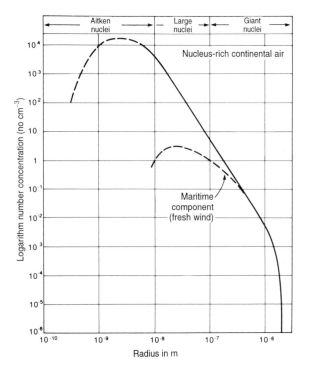

consider the physical and chemical influences these particles exert on climate.

3.5.1 *Aerosols, plankton, and climate*

The size distributions for aerosols in a marine, and heavily polluted terrestrial, atmosphere are shown in Fig. 3.12. There are millions of very small particles of less than 0.1 μm radius[2] in every cubic metre, but only a few thousand particles larger than 1 μm. The former are called Aitken nuclei, after their 1880 discoverer, while the latter are known as Giant nuclei. Those whose size lies between the two extreme sets are known as Large nuclei. The figure shows that the terrestrial atmosphere has far more Aitken and Large nuclei, but a similar number of Giants, to the marine atmosphere. The Giant nuclei are largely sea salt particles, whose significant impact on climate is discussed in the next section.

The Large and Aitken nuclei consist of a mixture of many minerals and chemical compounds. Our interest here lies in those with carbon, nitrogen, sulphur, phosphorus and iron content, as these contribute to biological processes if they are deposited in the sea. Iron is mostly to be found in aerosols derived from weathering of clays or other rocks. There is a small mid-oceanic aerial component, but most of this material is relatively large in size and likely to be lost through deposition in coastal zones. Atmospheric phosphate concentrations are also very low and not thought to make a significant contribution to oceanic levels.

Nitrogeneous compounds are particularly common in combustion and bacterial decay. The resulting aerosols can be small and while marine concentrations (0.1–0.4 μgkg^{-1}) are smaller than over land by an order of magnitude they are not negligible. Remote marine areas tend to have concentrations at the lower end of this range. Some of this nitrate will be from re-cycling of material of marine origin. However, seas close to shore, or exposed to air that has recently passed over continental areas, have enhanced concentrations which will contribute towards the supply of nitrate in the surface waters of the oceans. This is particularly true in the Northern Hemisphere, where man's activities are a dominating source of atmospheric nitrate and land masses occupy a much higher proportion of the surface area. Nitrate concentrations in waters subject to active algal growth will be of the same order as atmospheric levels, suggesting that aerial input could be of importance in supplying nitrate to fuel planktonic growth. It is believed, however, that this source is considerably less important than re-cycling of nutrients within the mixed layer. In §4.2.2 we will see that the gas nitrous oxide has a marine source as well as a terrestrial source.

About half of the sub-micron sized aerosols are composed of sulphates. Sulphate is an aqueous phase reaction product of the gas sulphur dioxide, SO_2, through the two reactions

$$2SO_2 + O_2 \rightleftharpoons 2SO_3$$
$$SO_3 + H_2O \rightleftharpoons H_2SO_4 \tag{3.11}$$

Sulpur dioxide has many terrestrial sources through combustion, volcanic emissions and biological activity. However a significant proportion of the atmospheric budget is the end result of biological activity in the ocean. This source, and the effect of sulphate on precipitation chemistry, will be discussed in §4.4.

---

[2] 1 μm is often called a micron.

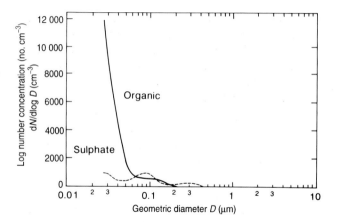

Fig. 3.13. Number size distributions of sulphate (dotted line) and organic aerosols (solid line) at an elevation of 1000m in an exposed location on Puerto Rico in the Caribbean. [Adapted from Fig. 1b of Novakov and Penner (1993). Adapted with permission from *Nature*, **365**, p. 824. Copyright (1993) Macmillan Magazines Limited.]

Another significant contributor to cloud condensation nuclei comes from a range of organic carbon-based aerosols. These contribute relatively little to the total aerosol mass, but their size distribution clusters below 0.05 μm and so they may have a greater impact on cloud droplet formation than their mass would suggest (Fig. 3.13). The particles are derived from direct emission from terrestrial combustion, or from gas-phase reactions of hydrocarbons within the atmosphere. Their contribution to the oceanic carbon budget is likely to be small, however. Their possible impact on cloud processes will be considered in the next section.

3.5.2 Sea spray, clouds, and climate

Sea salt is the largest contributor by mass to particulate material in the marine atmosphere. Between 10^9 and 10^{10} tonnes are cycled through the atmosphere each year. Residence times for particles vary from seconds to days, depending on the height to which the particles penetrate into the troposphere and the distance over which they are carried. Giant sea salt particles can travel a thousand kilometres; concentrations of such particles, chemically unaltered, can be almost unchanged hundreds of kilometres from the nearest ocean, provided no precipitation has occurred from the air mass in transit. Aerosols are returned to the Earth's surface by direct particulate, or dry, deposition, and through precipitation.

The latter process is known as wet deposition. The importance of sea salt for weather and climate was not appreciated until the 1940s when Woodcock began a series of pioneering projects, culminating in his presentation of the theory of sea salt-driven formation of raindrops in 1952. This theory is highly relevant to climate change and will be discussed in §3.5.3, taking into account more modern views on the importance of sub-micron, non-sea-salt particles for complete explanation of the observed droplet spectrum.

Over the oceans, in the planetary boundary layer, it is found that the concentration of salt particles in the atmosphere is strongly dependent on wind speed: the stronger the wind speed the more sea salt there is in the atmosphere. This is shown in Fig. 3.14, a summary of observations near

Fig. 3.14. Sea salt
concentrations as a
function of wind speed.
The upper (solid) line
shows the best-fit to some
of Woodcock's
observations near the
surface of the sea off Oahu,
Hawaii; the lower (dashed)
line shows earlier results of
Woodcock from
observations near the top
of the planetary boundary
layer.

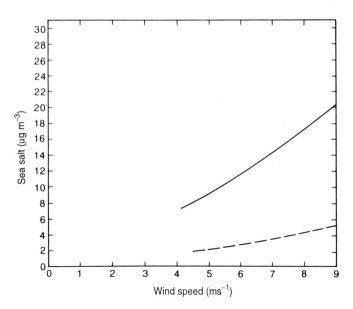

Fig. 3.14. Sea salt concentrations as a function of wind speed. The upper (solid) line shows the best-fit to some of Woodcock's observations near the surface of the sea off Oahu, Hawaii; the lower (dashed) line shows earlier results of Woodcock from observations near the top of the planetary boundary layer.

Hawaii. The size distribution of particles is also dependent on wind speed, although the tendency for many very light, but few heavy, particles is not. Fig. 3.15 illustrates this size–weight–wind speed relationship in the same experiment.

Within the boundary layer over the ocean there is often a sea salt inversion. At moderate wind speeds a rise in sea salt concentration occurs between 300 and 600 m. This peculiar phenomenon occurs both in the presence and absence of clouds. There are at least two possible mechanisms which might explain the inversion. Beneath clouds any fine drizzle often evaporates. This could leave the salt condensation nuclei at the evaporation level. Vertical wind shear could then allow the clouds to move faster than the sub-cloud air, leaving the salt-enriched air behind. Another mechanism involves the transport of sea salt from the surface into a more humid part of the boundary layer. Sea salt is hygroscopic, so the particles will grow in such an environment. Eventually some particles will become sufficiently large so that their *fall speed*, due to their mass, will be similar to the updraft raising them from the surface. Salt particles will then accumulate at this level of stability.

The hygroscopic nature of sea salt aerosol, its generally large size, and its abundance, are the principal factors giving this aerosol its climatic influence. Sea salt particles are large: greater than 0.1 μm, and very often 1 μm, in diameter. Aerosols visible through a microscope are predominantly composed of sea salt, which is largely sodium chloride, NaCl. At relative humidities greater than 75% NaCl particles become hygroscopic and attract water vapour to form small droplets. Other salts can become hygroscopic at much lower relative humidities – potassium carbonate, K_2CO_3, at 44%, for instance – but the abundance of NaCl makes this a significant hygroscopic aerosol.

Other, significantly smaller aerosols are also important in the formation

Fig. 3.15. The variation of
sea-salt particle cumulative
distributions with wind
speed, from Woodcock's
observations at the top of
the planetary boundary
layer.

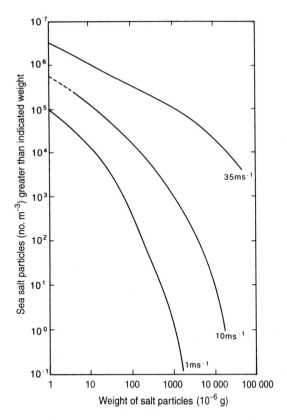

of cloud droplets. Sulphate particles, derived from biological or combustion sources (Chapter 4), form the major aerosol in the size range just below 1 micron radius (see Fig. 3.13)[3]. Even smaller sizes (below 0.1 micron) have been found be predominantly organic aerosols. Both of these source materials will actively contribute to cloud droplet formation, and are likely to dominate it, in terms of total number. Nevertheless, the larger droplets of sea salt origin have a crucial role in the precipitation process, as will be seen in §3.5.3.

Hygroscopic particles will make the atmosphere hazy, but it is only when the relative humidity reaches 100%, and the air becomes saturated with respect to water vapour, that clouds or fog appear. This is because hygroscopic particles, in absorbing water vapour, deplete the vapour concentration of the surrounding air. In any mix of particles this will mean that only a portion will be actively growing; the rest will have local microclimates with humidities below their threshold growth level. When the air is at saturation, or, as is common in clouds, slightly super-saturated, sufficient moisture is present for an optically dense mix of droplets to be

[3] Sulphate aerosols have several sources. Some, particularly of larger size, are derived from the sulphate salts of sea water spray (see Table 1.4). The remainder are termed non-sea-salt sulphate (or nss sulphate) aerosols. In general in this book, when sulphate aerosols are referred to they will be of nss origin.

generated. There will, however, still be a significant proportion of non-growing particles in such a mix.

Hygroscopic particles are essential for cloud formation in the atmosphere. It would require super-saturation humidities of 220–340% for spontaneous, or *homogeneous*, nucleation of water droplets from vapour alone. In providing a hygroscopic surface the aerosols allow *heteorogeneous* nucleation to occur in natural conditions. For small particles (less than 1 μm in diameter) such as nss sulphate or organic aerosols, and very marginal super-saturations, the radius of curvature of the forming droplet is important in determining growth. For larger, sea salt, droplets the rate of growth in the droplet radius, r, is a simple function of r and the level of super-saturation, S_s:

$$\frac{dr}{dt} = \frac{GS_s}{r} \tag{3.12}$$

In (3.12) G is an almost linearly increasing function of temperature. At 0°C $G = 6 \times 10^{-9}$ m^{2}s^{-1}, and 20°C $G = 1.23 \times 10^{-8}$ m^{2}s^{-1}. This simple formula is applicable to most cloud formation processes, where moisture is supplied by motion external to the cloud. For fogs, and possibly clouds forming through internal (radiative) cooling smaller droplets are more common and the radius of curvature of the droplets must be considered in studying the evolution of such clouds.

Condensation of water vapour around the larger aerosols is a necessary part of cloud formation leading to rainfall, as we shall see in §3.5.3. Because salt particles form the majority of the Giant condensation nuclei over the 70% of the globe that is covered by ocean (Table 2.2), such aerosols play a key role in the production of marine rainfall. Note, however, that the majority of the Large condensation nuclei over the ocean are sulphate particles (see Chapter 4). Over land there are many other sources of nuclei, from dust raised by the wind to particulates emitted by factories or formed by reactions within the terrestrial atmosphere. Cloud formation over the oceans requires strongly hygroscopic condensation nuclei, as the particle concentrations are rather lower than over continental interiors. Within marine clouds there are perhaps 20–200 droplets per cubic centimetre, while terrestrial clouds contain 200–2000.

Clouds can form for a variety of reasons; each involves bringing air to saturation. The most familiar perhaps is the ascent of moist air because of local convection, typically seen on a warm summer's day. Air warmed by a patch of ground hotter than its surroundings – perhaps a bare field, a city, or an island – becomes less dense and is forced to rise to seek its equilibrium density level. Rising air cools because it expands as it moves into regions of lower pressure. This parcel of air, sometimes known as a *thermal*, rises fast enough that there is little mixing with its surroundings. It can be shown that air rising without the gain or loss of heat, or adiabatically, loses temperature at a rate of 9.8°C/km. If the ascent is allowed to proceed unhindered the air will eventually reach saturation (see Fig. 2.5) and cloud formation will begin. Clouds formed in this way often display signs of the vigorous upward

motion by producing distinct puffs of cloud: these are characteristic of
cumulus clouds (Fig. 3.16).

Ascent of moist air can also be created by large scale dynamical
influences associated with fronts. This can occur over short distances,
giving rise to large cumulus clouds, or over hundreds of kilometres, when
layered clouds result. Air forced to rise over mountain ranges can also give
rise to cloud, if the mountains are high enough and there is sufficient
moisture content in the air.

Clouds can also be formed by cooling of the air *in situ*. This tends to
produce low level, layer clouds, but in unstable air low level cooling can
contribute to cumulus development when the cooled air mass passes over a
warmer surface. This happens if air flowing over a cold sea surface moves
over warmer land. The air has both cooled and gathered water vapour so
that when it is made more buoyant by heating from the warm land it can
rise to produce cumulus cloud. Such cloud formation can occur readily
around the North Sea or the west coast of North America in the spring,
when the land is warming but the sea surface is still cold.

Another mechanism for producing cloud, encountered in §2.10, is the
input of moisture from a warm sea surface into cold air. This can produce
low cloud or fog.

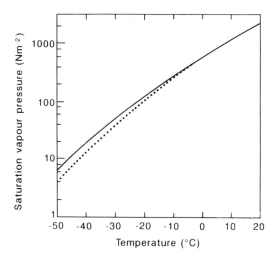

Fig. 3.17. The variation of saturation vapour pressure with temperature, over liquid water (solid line) and over ice (dotted line). Using data from Ludlam (1980).

In our discussion so far we have assumed that clouds are composed of water droplets alone. Ice clouds, or mixtures of water and ice, have been neglected. High clouds, or clouds at high latitudes, are, however, likely to be composed of ice crystals or snowflakes rather than water droplets. Aerosols that allow ice to *sublime* directly to the particle tend to be of terrestrial, rather than marine, origin; clays and heavy salts are most active as ice nuclei. Nevertheless, marine aerosols play a part in producing ice through the freezing of water droplets. Depending on the humidity and the size of the drop, water can exist in its liquid phase at sub-zero temperatures. In laboratory conditions tiny supercooled droplets can exist at temperatures of $-40°C$, but even in the atmosphere supercooling to below $-10°C$ is not uncommon. Once a droplet has been frozen by local cooling or through transport to a colder environment, the new ice crystal will begin to grow through sublimation. This is because the saturation vapour pressure for ice is lower than for water; air saturated with respect to an ice surface can be under-saturated with respect to water. This is shown in Fig. 3.17. In the upper troposphere the concentration of sea salt particles is drastically reduced from its boundary layer concentration. However, the necessity for saturation with respect to water, rather than ice, before cirrus clouds form suggests that instantaneous freezing of water droplets forming about salt, nss sulphate or organic carbon nuclei is an important mechanism for high cloud formation.

Stratospheric clouds, known to be important as sites for chlorine activation in the chemical process resulting in depletion of high level ozone, have sulphate particles as their main source of condensation nuclei. Most of these aerosols are thought to be of terrestrial origin, but as Chapter 4 will discuss, there are important marine sources of sulphate which may contribute to the stratospheric loading via entrainment of tropospheric air into the stratosphere in frontal regions or the ITCZ.

Clouds contribute to the climate in two ways. They play an important role in the radiational balance of the Earth. They are also the source for precipitation. We have encountered the radiational aspects of cloud physics

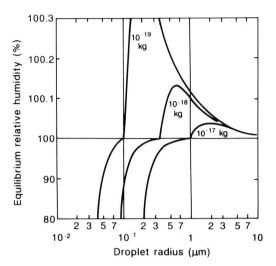

Fig. 3.18. The variation of equilibrium relative humidity with droplet radius, r, for a fixed mass of sodium chloride in water, for three different masses. Note that the scale has been expanded above 100% relative humidity to clarify the structure of within-cloud processes. [After Fig. 6.13 of McIlveen (1992). Reproduced with permission of Chapman and Hall from *Fundamentals of weather and climate* by R. McIlveen (1992).]

several times in earlier chapters. Clouds reflect, scatter, and absorb solar radiation. They also reflect, absorb, and re-radiate the Earth's radiation. Different types and thicknesses of clouds will affect the radiation balance in different ways. High clouds will have less water content, and thus will be optically 'thinner', allowing more *transmission* of solar radiation. They are also cold, thus emitting less infra-red radiation to space than a lower, warmer cloud. The cloud water content can have a double-edged contribution to climate: high water content makes a cloud a better solar reflector, but it will also act as a stronger greenhouse absorber. High water content also warms the cloud by releasing greater quantities of latent heat. Such warming is an example of *diabatic* heating. In Chapter 7 we will see how the reaction of cloud physics and chemistry to greenhouse gas emissions is a major area of uncertainty for the prediction of climatic change.

3.5.3   Mechanisms for precipitation formation

Precipitation occurs from clouds composed of water droplets or ice crystals. The previous section showed how, in a cloud of water droplets, differential growth rates will operate for different drops, depending on the local moisture content of the air and the radius of the drop. Very small droplets near larger droplets will have difficulty growing because their large radius of curvature increases the saturation vapour pressure compared to that for larger drops. This is illustrated in Fig. 3.18. Once a few larger drops form, therefore, they will scavenge vapour from nearby small droplets by forcing the latter's local humidity below saturation. Even if there is competition for vapour between two relatively large drops, the larger will capture more vapour. This is because the rate of change of volume, V, for large drop growth is a linear function of radius:

$$\frac{dV}{dt} = 4\pi r G S_s \qquad (3.13)$$

Fig. 3.19. Schematic
illustration of the
coalescence process.

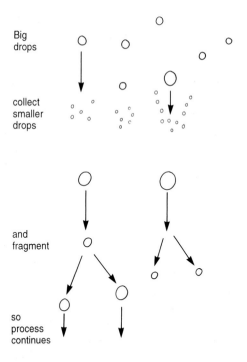

Note that the typical distance between cloud droplets may be several hundred droplet radii, that is 0.1–1 mm.

The larger drops will eventually become heavy enough for their gravitationally-induced fall speed to become greater than the ascending motion within the cloud. For example, a drop of radius 10 μm has a fall speed of 1 cms^{-1}. As these drops fall through the cloud they will grow by sweeping up smaller drops in their path. If they gain enough mass the drops will become unstable and splinter into a number of smaller, but still relatively large, drops. These will continue to descend, capturing more and more water, perhaps undergoing repeated shattering, until they fall out of the cloud. This process of *coalescence* is shown schematically in Fig. 3.19. The resulting raindrops will then reach the ground as precipitation, unless evaporation below the cloud base converts the drops back to vapour before impact. In the latter case an observer will see trails of rain seemingly suspended beneath the cloud in what are known as *virga*. The time required for rain to form depends on the initial size distribution of the droplets and the water content of the cloud. The fastest rates of 10–30 minutes occur in very moist air with few condensation nuclei – and hence large droplets. Large numbers of nuclei with relatively small moisture levels lead to clouds which will never reach the stage of precipitating. The warm marine atmosphere of the tropics is ideal for potential precipitation.

Clouds in the winter mid-latitudes and polar regions will be a mix of a few ice particles and a large number of tiny cloud droplets. We have seen that this is an unstable situation because of the saturation vapour pressure difference over ice and water causing the ice particles to grow preferentially. Eventually the large ice crystals will become large enough to precipitate as snow, or rain if the temperature of the lower atmosphere is somewhat above freezing. This is known as the *Bergeron–Findeisen* process of rain formation, after the Norwegian scientists who proposed the mechanism in the 1930s. At one time it was thought to be the major mechanism for rain production, but it is now realised that the coalescence process is more widespread. Even within cold clouds, where the Bergeron–Findeisen process is operating, collisions between ice crystals and water droplets add mass to the crystals much faster than preferential vapour attraction. The Bergeron process thus requires a moist atmosphere with adequate supplies of ice and water nuclei. Continental margins would seem to be the most likely areas for these conditions to be found.

3.6 Photochemical reactions in sea water

There are innumerable atmospheric photochemical reactions in which molecules are fragmented, or given different reactivity by changing their energetic state. In Chapter 1 we encountered the climatically important *photolytic* formation of ozone in the stratosphere (equation 1.4). In Chapter 4 we will see how a photolytic reaction removes hydrogen sulphide from the atmosphere rapidly, and also how sulphur dioxide is photolytically oxidised in forming sulphate aerosols. Reactions driven by the energy of a photon of appropriate energy are also important in the production of smogs over industrialised cities.

In the ocean the possibility of photolytic reaction is often neglected, yet Chapter 4 describes the consequences of just one: photosynthesis. In the right conditions appreciable amounts of solar radiation can penetrate tens of metres beneath the surface of the sea (Fig. 2.2). Sea water preferentially absorbs the longer, less energetic, red wavelengths of the visible part of the solar spectrum. Thus the shorter, high energy, visible wavelengths provide much of this penetrative power. Ultra-violet light, however, is also strongly absorbed and scattered. In the atmosphere ultra-violet light is a potent driver of many photochemical reactions; most of the solar photons in this energy band have been absorbed in the atmosphere before the solar radiation reaches the ocean surface; what remains is rapidly removed in the first few metres upon penetration.

Chemical reactions driven by photons can take a number of different forms. One of the most important is *photo-dissociation*: a fragmentation of the chemical *species* upon interaction with a photon. The photon must have sufficient energy to break the chemical bond; photons with greater energy additionally impart kinetic energy to the resulting fragments. One of these fragments is typically in an excited state, that is, its electrons will occupy *orbitals* above the atom's (lowest energy) standard *electron configuration* (see Appendix B). A special case of this type of reaction is known as *photo-ionization*. This occurs if the fragments from the reaction are not discrete atoms or molecules but merely an electron and the reacting atom's or molecule's positively charged ion.

Absorption of a photon can lead to the specie radiating energy in order to return to its standard energy state. This is called *luminescence*, but

Table 3.4. *Climatically important photochemical reactions in sea water*

$NO_2^- + H_2O + v \rightarrow NO + OH + OH^-$
$NO_3^- + v \rightarrow NO_2^- + O$
$CH_3I + Cl^- + v \rightarrow CH_3Cl + I^-$
$CO + v \rightarrow CO^*$ (an excited, more reactive state of CO)
$H_2S + OH^- + v \rightarrow H_2O + HS$
$(CH_3)_2S + 5O_2 + v \rightarrow 2CO_2 + 2H_2O + H_2SO_4$

particular types of this reaction are known as *fluorescence* or *phosphorescence*, depending on the type of energy transition. Such a reaction concentrates radiation from a wide spectrum of incident wavelengths to one, making this one emission wavelength particularly intense. Both of these reactions are seen in sea water; fluorescence can be used as a *chlorophyll* tracer, chlorophyll being one of the products of photosynthesis (see §4.1.3).

Photons absorbed by a specie do not always result in fragmentation or luminescence. The raised energy state of the specie can be relatively stable, giving it a different reactivity and chemistry. The energy can also be given to other molecules via collision, converting the energy into kinetic energy, molecular vibrations, or transferring an excited state to the colliding molecule. Excited molecules or atoms often have greatly enhanced reactivity, so that they react with other molecules very rapidly.

There are a number of photochemical reactions in the upper layers of the ocean that influence the concentrations of climatically sensitive gases, in addition to those mentioned at the beginning of this section. Nitrate, NO_3^-, and nitrite, NO_2^-, two nutrients important for phytoplankton growth, can be photolysed to produce different oxidation states. A number of trace gases produced in the ocean, such as methyl iodide, CH_3I, dimethyl sulphide, $(CH_3)_2S$, carbon monoxide, CO, and hydrogen sulphide, H_2S, will all undergo photochemical transformation in sea water as well as the atmosphere. A summary of presently known reactions is given in Table 3.4. This is a field of growing interest and new, climatically important, reactions are likely to be added in the future, for instance the photochemistry of organic iron colloids (see §4.1.1).

These photochemical reactions complicate the estimation of fluxes of gases to the atmosphere, as they introduce a diurnal cycle that may enhance or counter any similar biologically driven cycle. Observational study is as yet insufficient to determine the importance of the diurnal cycle in many air–sea gas exchanges.

This brief discussion of photochemistry provides useful background when considering biological processes in Chapter 4 and the atmospheric photolytic reactions important in the stratospheric ozone problem in Chapter 7.

3.7   Chemical tracers

Oxygen was seen in §3.4 to be useful as a tracer of deep water movement. Other chemicals can also be used to trace water movement, either because

Fig. 3.20. The distribution of $^{14}C/C$ ratios at a depth of 4000m in the main basins of the global ocean. The data is expressed as the part-per-thousand difference from the $^{14}C/C$ ratio in the atmosphere prior to the onset of industrialisation and normalised to a constant $^{13}C/C$ ratio. [Fig. 5.4 of Broecker and Peng (1982). Reproduced with permission of W. S. Broecker.]

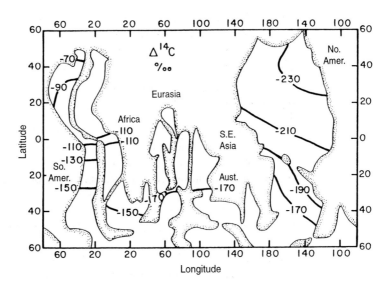

the compounds concerned are inert, or because reaction, or radioactive decay, of the compound is known, at least qualitatively. These tracers have various sources: from the atmosphere, within the water column, or from the sea floor. The radioactive isotope of carbon, ^{14}C, for instance, shows a similar pattern horizontally to oxygen. Fig. 3.20 shows the ratio of the ^{14}C to the most common carbon isotope, ^{12}C, at a depth of 4000 m around the world ocean. This radioactive isotope is formed in the atmosphere by the reaction of a cosmic ray from beyond the Earth with nitrogen atoms in the atmosphere. The radioactive decay of ^{14}C through the emission of an electron (e^-, or a β particle),

$$^{14}C \rightarrow e^- + {}^{14}N \tag{3.14}$$

with a half life of 5700 years, means that once carbon, and hence ^{14}C atoms, enter the sea there is no new source of ^{14}C and so the proportion of the unstable isotope declines as the water mass originally in contact with the air is subducted. Fig. 3.20 therefore depicts the movement of water from the North Atlantic and Weddell Sea into the deep regions of the entire ocean. This tracer suggests that the northeast Pacific contains the oldest water and that it has taken approximately a thousand years for the water to reach this location since it was last at the surface.

Apart from oxygen and ^{14}C the principal oceanic tracers of atmospheric origin with which we shall be concerned here are anthropogenic and have only been added to the atmosphere, and therefore the ocean, in quantity since about 1940. These tracers can be used to track both where water is being subducted from the surface, and also where it travels once it sinks. The tracers of concern are *freons*, or chlorofluorocarbons, tritium and helium. The first of these are used extensively in industry as refrigerants, propellants and solvents, although their use is being phased out as part of the response to the ozone problem (see §7.3). Tritium, a radioactive isotope

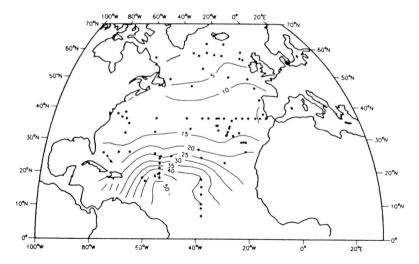

Fig. 3.21. The tritium–helium age, in years, on an isopycnal that surfaces in the northern Atlantic but is below the thermocline in the sub-tropical North Atlantic. [Fig. 4a of Jenkins (1988). Reproduced with permission of The Royal Society from W. J. Jenkins, *Phil Trans. Roy. Soc. London*, **A325**, 1988, 43-59.]

of hydrogen (^3H), is a product of atmospheric testing of nuclear weapons. Its decay product, through the same process as reaction (3.14),

$$^3H \rightarrow {}^3He + e^- \tag{3.15}$$

is helium, and because of the relatively short life time of tritium (12.26 years) the combined tritium and helium concentration can be used to determine the age of the water since it was last *ventilated* at the surface (see §3.1 for a discussion of how tritium enters the ocean).

Use of freons as tracers is still fairly novel, and hampered by the large volumes of water required to extract experimentally detectable amounts of these gases. The principal technique for tracking water masses with these tracers stems from the different emission histories of the freon gases. The principal two that have been used industrially are Freon-11 ($CFCl_3$) and Freon-12 (CF_2Cl_2). From knowing their historical atmospheric concentrations (which are mixed throughout the troposphere globally in a few months) and examining the ratios of CFC-11/CFC-12 the age and spreading of a water mass can be determined. These gases have no natural sources to bias the observations.

The use of tritium/helium is somewhat similar to freons. It has been used as a chemical tracer since the 1970s. The atmospheric concentration, which has a geographical bias due to the location of hydrogen bomb tests and the limited time over which they occurred (late 1950s–1960s), is known. Therefore the ratio of tritium to the total tritium and helium level can be used to find the age, and spreading, of water in the intermediate and deep water. This tracer is complicated by the existence of helium in both the atmosphere and ocean naturally. The atmospheric concentration of helium is extremely low, as can be inferred from its absence from Table 1.1, but it will contribute a weak background signal. Also, helium is released from areas of geothermal activity under water, particularly along oceanic ridges. This adds to the background, and will be particularly handicapping for study of very deep water movement. Therefore the helium measured in a tracer sample has to be corrected for these background levels. Fig. 3.21 shows the tritium–helium age for an isopycnal surface in the main thermocline of the North Atlantic Ocean. This shows that water requires about 10 years to move half way around the sub-tropical gyre.

Another, non-anthropogenic, tracer which has recently been used to provide information about oceanic flow patterns is the ratio of the rare isotope of oxygen, ^{18}O, to the common isotope, ^{16}O, in ordinary sea water. ^{18}O is not radioactive, but as we will see in §6.1.1, there is a temperature-dependence of the ^{18}O:^{16}O ratio in sea water ultimately related to preferential evaporation of ^{16}O at the expense of ^{18}O. The ^{18}O abundance in sub-surface sea water can therefore be related to its surface origin. This concept has been used to suggest that Antarctic Bottom Water formed in the Weddell Sea may have two sources with similar temperature and salinity: one close to the shelf ice edge, where melting of very ^{18}O-depleted glacial ice has lowered the local sea water's ^{18}O:^{16}O ratio, and another further off-shore at the sea-ice edge, where more characteristic oceanic ^{18}O:^{16}O ratios for these latitudes are found.

Further reading

A complete reference list is available at the end of the book but the following is a selection of the best books or articles to follow up particular topics within this chapter. Full details of each reference are to be found in the Bibliography.

Broecker and Peng (1982): An invaluable guide to ocean chemistry. Well written with a very comprehensive list of pre-1982 references. Also a very good guide to the chemical tracers section.

Ludlam (1980): A comprehensive book on cloud processes.

McIlveen (1992): A very readable section on cloud microphysics.

Open University Oceanography Series (1989): The Sea Water and Ocean Chemistry volumes of this series give an excellent introduction to the properties of sea water, general ocean chemistry, and the links between chemistry and ocean sediments.

Pinet (1992): A general oceanography text with good introductory material on oceanic chemistry.

Rogers and Yau (1991): A more modern text for the cloud microphysicist.

4 Biochemical interaction of the atmosphere and ocean

The oceans teem with life on all scales from microscopic plant and animal life to blue whales. Tiny plants known as *phytoplankton* form the base of the numerous intricate food chains. These organisms photosynthesise; they are therefore found in the surface layers of the ocean. Most biological activity in the ocean also occurs in this region, although marine organisms exist at all depths, including within the sediments on the sea floor. Phytoplankton, and other marine organisms, constitute a component of the carbon cycle (§3.3) and produce chemical compounds that ultimately influence the concentrations of particular atmospheric gases and aerosols. The influence of marine biology on climate, through such mechanisms, has only recently been appreciated; this chapter attempts to discuss the biological processes which may influence the climate, and assesses their importance.

4.1 Phytoplankton

Plankton are defined as living organisms within the sea which are essentially restricted to moving with the prevailing current; they are not, however, completely devoid of motility, as some species swim upwards towards stronger light conditions during daylight hours while others move deeper to regions of lower intensity. Phytoplankton are plant species within this category of organisms. They are so called because they photosynthesise - hence 'phyto'plankton. Commonly, there are a large number of different species within any one population. Differences in species' biological behaviour can be important in discussions of population dynamics, and will appear in §§4.3 and 4.4 when we discuss chemicals produced by plankton that influence the climate, but in this general description they will usually be ignored.

Phytoplankton are very important in the control of atmospheric carbon dioxide via oceanic uptake. As we saw in §3.3.2 the use of CO_2 in photosynthesis creates large air–sea gradients in p_{CO_2}, drawing much more of the gas into the ocean than would otherwise occur. This additional draw-down is variable in space and time, but contributes significantly to the modification of anthropogenic atmospheric CO_2 input, as will be seen in §7.2.4, and may be involved with past climatic change, as will be discussed in Chapter 6.

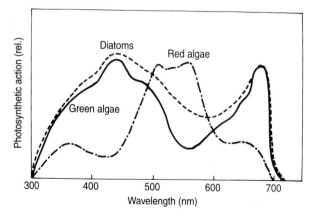

Fig. 4.1. Relative absorption spectra for three major marine algal groups: diatoms, green algae and red algae. [Reprinted from Parsons *et al.*, *Biological oceanographic processes*, Copyright 1984, 330 pp., with kind permission from Elsevier Science Ltd., The Boulevard, Langford Lane, Kidlington OX5 1GB.]

4.1.1   *Phytoplankton growth*

Phytoplankton photosynthesise to produce carbohydrates. Other organic compounds, such as proteins and lipids, can also be formed by photosynthesis. The proportioning of products depends on environmental conditions and the plankton's age. This complex process can be summarised by

$$nCO_2 + 2nH_2O \underset{\text{respiration}}{\overset{\text{visible light}}{\rightleftharpoons}} nO_2 + nH_2O + n(CH_2O) \qquad (4.1)$$

Different species preferentially absorb different wavelengths of visible light during this process. The absorption spectra for three broad types of phytoplankton are shown in Fig. 4.1. Note that the algae denoted by a colour tend *not* to absorb that wavelength band.

In §2.1.1 we saw how solar radiation is rapidly absorbed within the ocean so that little such energy is left at a depth of 100 m. The penetrative power of solar radiation is also a function of the water clarity, as shown in Fig. 2.2. This will be a function of location. We further saw in that section that red light is preferentially absorbed by sea water, allowing the blue–green wavelengths to penetrate deeper. Too much radiation also limits photosynthesis. The zone of maximum photosynthesis is thus below the surface, where there is still sufficient radiant energy of the appropriate wavelengths, but not so much that reaction (4.1) is suppressed. These characteristics, and the annual and daily variation in the intensity of the phytoplanktons' energy source, will strongly control the temporal, vertical and geographical distribution and abundance of phytoplankton.

There are other controls on phytoplankton abundance. Photosynthesis provides the phytoplankton with energy but the basic building block of the physical structure of the plant's cells is nitrogen. Nitrogen as a gas, N_2, can be fixed by some species of algae, particularly fresh-water species. The blue–green algae that are common in British waterways during hot, dry summers are one such species. However, oceanic species have to rely on absorption through cell walls of nitrogeneous compounds, particularly nitrate, NO_3^-, nitrite, NO_2^- and ammonium, NH_4^+.

Phosphorus is another control on phytoplankton growth. It is also a component of the cell structure, but is more important as a constituent of

Fig. 4.2. Typical seasonal
cycle in phytoplankton
(solid line) and
zooplankton (dashed line)
communities in the North
Atlantic. [Reprinted from
Parsons *et al.*, *Biological
oceanographic processes*,
Copyright 1984, 330 pp.,
with kind permission from
Elsevier Science Ltd., The
Boulevard, Langford Lane,
Kidlington OX5 1GB.]

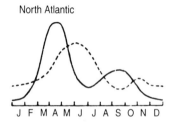

North Atlantic

J F M A M J J A S O N D

the cell's genetic make-up, or DNA. It thus participates in cell reactions. Phosphorus is mostly absorbed by the organism as the phosphate ion, PO_4^{2-}.

Nitrate and phosphate are known as limiting nutrients for phytoplankton. Even in good lighting conditions there must be adequate supplies of these compounds for phytoplankton growth to occur. Plankton growth shows a strong seasonal cycle away from the tropics, illustrated in Fig. 4.2 (see, however, §4.1.2). This is due to the interaction between light intensity and nutrient supply. During the winter there are low light levels, but the mixed layer is deepened to mix supplies of nutrient from the thermocline into the upper ocean. As spring begins light levels increase and the mixed layer shallows because of the onset of thermal stratification. There is fresh nutrient available and a riot of growth, known as the *spring bloom*, occurs. This quickly uses up nutrient, and also allows *zooplankton* – animal plankton which feed on the phytoplankton – to dramatically increase in population. For some months a balance is maintained between nutrient supply, phytoplankton growth, and predation by the zooplankton and other marine life-forms. There may be a second, smaller, bloom in the autumn as deepening of the mixed layer begins, while light levels are still conducive to active photosynthesis. To maintain productivity during the summer a mechanism for re-cycling nutrients is required, as only limited diffusion from beneath the thermocline, or from atmospheric input, replenishes the reservoir of nutrients. Bacteria provide this mechanism by decomposing dead organisms, or the excreta of living organisms, so that their cells become available as nutrients for later generations. The bacteria, of which a number of species are required to fully decompose any matter, excrete enzymes that react chemically with the *detrital* material. The process of bacterially-induced decomposition is called bacterial *demineralization*. This cycle is shown schematically in Fig. 4.3. The speed of conversion of nitrogenous or carboniferous material from waste through decomposition to re-incorporation into new organisms is very variable. It can occur over a few hours to days, or take several thousand years. During spring blooms this conversion seems to be particularly fast.

There are other factors important in phytoplankton growth such as temperature, salinity, pH, availability of iron, carbon and silicon. Iron and silicon may act as limiting factors for particular species. The iron required by phytoplankton cannot be assimilated from its commonest state in oceanic waters, as part of particles or *colloids*. It is obtained from dissolved

Fig. 4.3. Schematic illustration of the cycling of nutrients within the mixed layer.

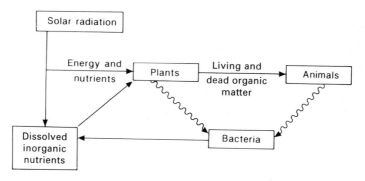

iron compounds involving the hydroxyl or OH^- ion, or *radical*.[1] It has been proposed that Southern Ocean productivity is limited by iron deficiency, but more recent work suggests that nutrient limitation is perhaps more important and that even assured availability of iron would not increase productivity sufficiently to impact on the oceanic uptake of carbon dioxide. The dissolution of iron from its colloidal form is enhanced by photolysis involving light with blue–green wavelengths (300–400 nm). Iron may therefore be made available by this process for plankton uptake deep in the euphotic zone.

Silicon functions in a similar way to nitrogen for some species, for instance diatoms. Silicate (SiO_4^{2-}) is abundant in coastal waters because it comes from dissolved clays in run-off. Diatoms are thus often restricted to these areas, or appear in number in much of the ocean only during the spring bloom, when the winter supply of entrained silicate is still relatively abundant.

The oceanic concentration of the basic building block of the phytoplankton cells – carbon – is two or three orders of magnitude greater than nitrate or phosphate. Carbon dioxide, through photosynthesis, is vital for this conversion. Atmospheric levels of CO_2 seem high enough not to limit phytoplankton growth but laboratory-based experiments have indicated that sufficiently low oceanic partial pressures may hinder cell growth. Biological productivity may therefore have been CO_2-limited in some areas under the significantly lower atmospheric carbon dioxide levels of glacial times (~ 180 ppm – parts per million). The interaction of atmospheric CO_2 and marine productivity in glacial climates will be considerd further in §6.2.1.

Notwithstanding this discussion, however, the limiting factors for the growth of most species and regions are the supply of nutrients and light, while temperature acts as a regulator of the speed of growth.

4.1.2 Geographical variation

The growth cycle of phytoplankton given in Fig. 4.2 is very much a North Atlantic picture. In fact there are radically different seasonal cycles in

---

[1] These compounds contain FeIII, or iron in compounds where the oxidation state is such that there are three less electrons associated with the iron atom than in the elemental atom.

Fig. 4.4. Summary of
seasonal cycles in plankton
communities in different
parts of the world.
Phytoplankton biomass is
shown by a solid line,
zooplankton with a dashed
line. The scales are relative
and not necessarily
comparable from one site
to another. [Reprinted
from Parsons *et al.*,
*Biological oceanographic
processes*, Copyright 1984,
330 pp., with kind
permission from Elsevier
Science Ltd., The
Boulevard, Langford Lane,
Kidlington OX5 1GB.]

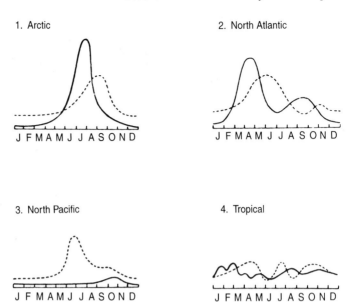

different parts of the world ocean, depending on the background light, nutrient and temperature patterns. Four different regimes are shown in Fig. 4.4, for polar waters, North Atlantic, North Pacific and tropical areas. In the polar oceans there is only sufficient light for one bloom during the summer. This occurs rather later in the season than the spring bloom of the North Atlantic double-bloom cycle. In the North Pacific, zooplankton of the species *Calanus plumchrus* and *Calanus cristatus* hatch from eggs laid by adults in late winter just as conditions become propitious for phytoplankton growth. No bloom is seen because these young zooplankton consume any new growth. Late in the summer the zooplankton population may decline sufficiently for a small autumn bloom in phytoplankton. In the tropics there is sufficient light and heating throughout the year, so that there are continual small rises and falls of population as zooplankton and phytoplankton populations, and nutrient levels, interact.

There are also changes from year to year in plankton population. These can be linked to changes in the climate, through winter mixing variability, temperature and light availability. For instance, during the cooling of western Europe in the 1960s (see §6.4.5) there was a marked reduction in the size of phytoplankton blooms in the North Sea, as well as a shortening of the growth season. Large variations are also experienced in tropical Pacific *primary productivity* during El Niño events (§5.2). There may also be changes in the supply of nutrients from land run-off. Disease and fishing practices may also alter the balances within the ocean that affect phytoplankton stocks.

The total phytoplankton mass produced per year within the world ocean is about 31×10^9 tonnes of carbon. The geographical variation in this is strongly tied to the shape of the ocean basins. As can be seen in Fig. 4.5, most of the ocean's production occurs in coastal zones, where the supply of nutrients is continually renewed from land sources. We saw in §3.5 that

Fig. 4.5. Distribution of primary production in the World Ocean. [Reprinted from Parsons *et al.*, *Biological oceanographic processes*, Copyright 1984, 330 pp., with kind permission from Elsevier Science Ltd., The Boulevard, Langford Lane, Kidlington OX5 1GB.]

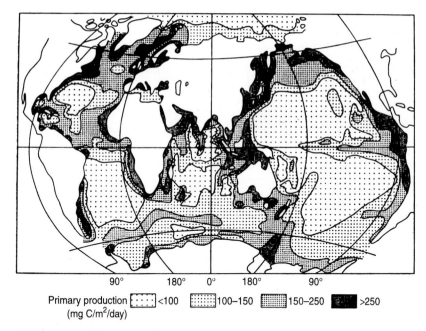

Primary production ⬚ <100 ⬚ 100–150 ⬚ 150–250 ■ >250
(mg C/m^2/day)

airborne nutrient deposition into the sea is negligible. Some coastal production, however, is associated with upwelling zones, where nutrients are supplied from beneath the thermocline. Such regions are off west Africa, Arabia, and the western coast of South America. In §2.10.2 we discussed the physical processes responsible for this upwelling.

The centres of the sub-tropical gyres are clearly minima of production. This occurs because of the basic downwelling that is induced by Ekman convergence within these gyres, as discussed in §2.10.1. Tropical regions also tend to have low productivity, as Fig. 4.4 suggests, because of the intense and continual competition between zooplankton and phytoplankton; there is also little entrainment of nutrients from below the mixed layer. Exceptions to this are found near the equator, particularly in the Atlantic and eastern Pacific. The upwelling induced along the equator by Ekman divergence, discussed in §2.10.3, brings supplies of nutrient and carbon dioxide to the surface allowing greater productivity.

There are other regions of the ocean where productivity seems higher than might be expected. The northern Atlantic and Pacific, the Atlantic sector of the Southern Ocean, and southeast of Australia all have elevated levels of production. This is partly because these areas of ocean are away from the centres of the sub-tropical gyres, but they are also regions of active eddy formation in vigorous currents. When cold, cyclonic eddies are produced by the instabilities in these currents the equatorward movement of the nutrient-rich surface waters, allows enhanced productivity. Such eddies can sustain enhanced production for many months, producing local phytoplankton growth in the middle of the North Atlantic gyre, for instance. Western boundary currents, and the Antarctic Circumpolar Current are prime locations for copious eddy formation. It should also be noted, however, that warm, anticyclonic eddies have the opposite effect.

Fig. 4.6. Abundance of the
zooplankton *Sagitta
pseudoserratodentata* along
a three kilometre track.
Note the large- and
small-scale structure in the
patchiness. The arrow
denotes the mean
along-track abundance.
After Parsons *et al.* (1984).

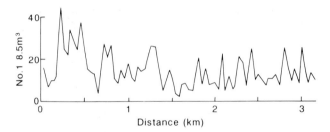

Their downwelling flow tends to decrease productivity by stopping entrainment of deep supplies of nutrients.

Enhanced productivity can also occur in the sea beneath intense atmospheric low pressure systems. These can generate local cyclonic eddies within the water, producing upwelling and sparking a local phytoplankton bloom. This effect is generally relatively limited, however, as deep penetration of mixing needs thermodynamic, as well as wind, forcing.

Superimposed on these larger scale geographical variations are dramatic changes over small distances. Phytoplankton respond to the local conditions, as well as the wider constraints of the general circulation. Thus populations can be very patchy. Fig. 4.6 shows variation in a zooplankton species, *Sagitta pseudoserratodentata*, over a 3 km distance, revealing changes in population size approaching an order of magnitude. There can also be sharp variations in production near oceanic fronts, separating water of distinctly different characteristics, as these zones tend to be regions of convergence. Langmuir circulation cells are another small-scale feature that can be associated with local productivity variability, as we saw in §2.9.2. Full description of planktonic activity in the ocean is therefore an extremely difficult task and beyond current technology, even using satellite-based instruments discussed in the next section.

4.1.3   *Vertical variation and ocean colour*

Primary production is often traced by measuring the chlorophyll concentration, as this chemical is associated with photosynthesis and is well-correlated with plant biomass. Peaks in chlorophyll can be found both near the surface, and at significant depth. However, a chlorophyll maximum generally occurs near the base of the euphotic zone. This is in association with the *nitracline*, the region where nitrate concentrations begin to rapidly increase with depth from the surface minimum.

There are a number of factors contributing to this productivity. The light levels are often less than 10% of those at the surface, but nutrients are more abundant in this region. Also, there is less vertical motion in the water column near the base of the mixed layer, allowing the plankton more time to take up nutrients. In §4.1.1 it was seen that accessible iron may be more readily available through photolytic reactions at these depths.

The depth of this chlorophyll maximum is variable, being as deep as 100–250 m in the sub-tropical gyres and as shallow as 20 m in coastal seas. This variation is dependent on the clarity of the water, as well as the

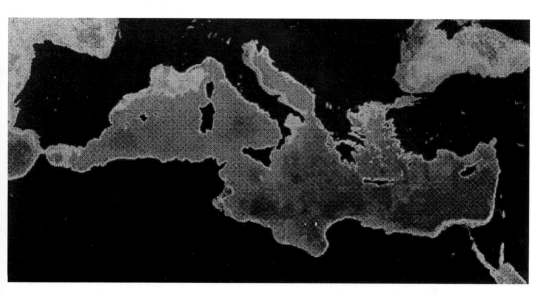

Fig. 4.7. A CZCS image of the Mediterranean Sea. Lighter shading indicates more productivity. Note the strong coastal bias and the eddy motion south of Spain and France.

circulation patterns. There is also a feedback between phytoplankton population and this chlorophyll level. The more production generated the less light penetrates to deep water, because of the shading effect of the organisms. Highly productive regions thus have relatively shallow chlorophyll maxima. This will affect the heat balance of the upper ocean by allowing more or less radiational energy to penetrate to a given depth.

The concentration, and vertical variation, of phytoplankton thus affects the light transmissivity of the ocean. The absorption of radiation with depth will not only be dependent on production, but also the colour of the ocean with respect to the atmosphere. We have seen, in §2.1.1, that sea water preferentially absorbs at the red end of the solar spectrum. In the absence of algae the water will appear blue. Phytoplankton tend to absorb wavelengths at both ends of the spectrum, leaving the green band to be reflected; hence the green colour of chlorophyll in the sea (and in leaves). Primary production will therefore tend to make the sea appear green, rather than blue. Coastal waters are often a rich mix of production and sediment, becoming in consequence a murky grey colour.

This variation in the properties of light reflected from the ocean surface can be used by satellites in estimating the amount of production occurring in the oceans beneath them. The Coastal Zone Colour Scanner (CZCS), an image from which is shown in Fig. 4.7, was an instrument aboard NASA's NIMBUS–7 satellite, and was in orbit in the late 1970s and 1980s. It has provided considerable information about the distribution of productivity over the world's oceans.

The variation in the wavelength characteristics of reflected light from the oceans due to changes in biological productivity will also have a climatic influence. Pushing the predominant wavelength of reflected light from blue to green, if productivity increased, could lead to a greater reflection of energy from the ocean surface, because there is more energy in the green band of the solar spectrum (Fig. 1.2). Regions of productivity would therefore reflect more incident radiation, and also allow less of it to warm the lower reaches of the upper ocean. A contrasting effect of phytoplankton is the warming of the upper ocean by the increased absorption of radiation as biomass increases. Quantitative estimates of ocean productivity derived from satellite remote sensing are, however, complicated by the unsolved problem of how to extrapolate the observed surface signature to the peak productivity at depth. In 1996 it is planned that a new instrument for monitoring ocean productivity (SeaWIFS) will be launched on the European Space Agency's ERS–2 satellite, allowing a resumption of such scrutiny of the global ocean.

4.2   Climatically active products of marine biological processes

Carbon dioxide is a major participant in marine biological processes, as well as being an important greenhouse gas. It has been discussed in several places already, and will be further considered in the next section. There are a number of other by- products of marine biological activity which are active in the climate: these include several carbon-, nitrogen-, and sulphur- based gases, methyl chloride and methyl iodide. Estimates of the global fluxes of some of these, and other, gases across the air–sea interface are given in Table 3.2.

4.2.1   *Carbon compounds other than CO_2*

Methane, CH_4, and carbon monoxide, CO, are the principal two climatically active carbon gases, other than CO_2, produced by biological activity in the oceans. They are both greenhouse gases of some significance (Table 1.2).

Methane is a product of anaerobic decay, that is bacterial decay in the absence of oxygen, for example:

$$2CH_2O \xrightarrow{\text{bacteria}} CH_4 + CO_2 \qquad (4.2)$$

There are several other chemical pathways by which alcohols or carbon dioxide may be reduced to methane. The ocean contributes only 1–2% of the total methane input to the atmosphere. Oceanic methane production may be associated with regions of high phytoplankton biomass. The oxygen minimum below the surface waters (§3.4) due to biological utilisation may then be sufficiently severe that the *reducing* environment necessary for methane-producing decomposition exists. Such a situation, encountered in the Arabian Sea in September 1986, is shown in Fig. 4.8.

There is a large flux of carbon monoxide to the atmosphere from the ocean (Table 3.2). Some of this will have been formed by photochemical oxidation of methane (see §3.6) but much of it is produced by microbiological activity. Carbon monoxide is a product of incomplete respiration, that is, oxidation when there is an inadequate oxygen supply for complete respiration to produce CO_2, for example:

$$HCO^- + O_2 + M \rightarrow CO + HOO^- + M \qquad (4.3)$$

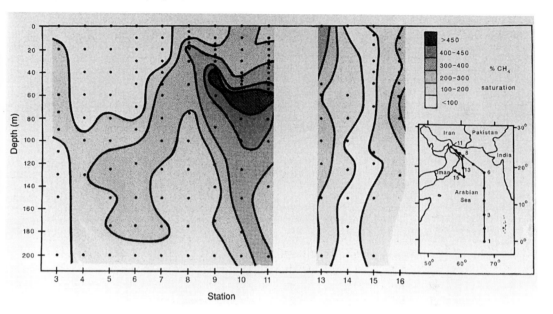

Fig. 4.8. The percentage saturation of dissolved methane, with depth, along a transect through the Arabian Sea at the end of the summer monsoon period. Stations 7–16 are in part of the upwelling region off the Arabian coast at this time of year. [Reprinted with permission from *Nature*, **354**, Owens *et al.*, pp. 293–5. Copyright 1991 Macmillan Magazines Limited.]

4.2.2   Nitrogeneous compounds

Nitrous oxide, N_2O, and ammonia, NH_3, are the principal nitrogen-based gases given off by marine biological processes. During the oxidation of organic material by phytoplankton some of the nitrogen is converted to N_2O rather than nitrate, that is,

$$2O_2 + 2NH_4{}^+ \rightarrow N_2O + 3H_2O + 2H^+ \tag{4.4}$$

rather than

$$2O_2 + NH_4{}^+ \rightarrow NO_3{}^- + 2H^+ + H_2O \tag{4.5}$$

This is seen in Fig. 4.9, which shows a clear peak in concentration at the oxygen minimum. The conversion rate is small; only one N atom in 1000 is converted to N_2O rather than nitrate. In some areas of the ocean, where the oxygen level can be taken to zero by biological uptake, nitrate and nitrous oxide become the source of oxygen and so some of the gas can be re-cycled within the ocean. The northwest Indian Ocean, the sub-tropical east Pacific, and the deep waters of the Bering Sea show this behaviour.

Nitrous oxide is also produced on land as a by-product of combustion and aerobic bacterial activity in soils. The ocean is a source of N_2O, however; it contributes about a quarter of the net input to the atmosphere each year. Nitrous oxide is a minor, but increasing, greenhouse gas – see Table 1.2 and §7.2.1.

Ammonia is produced by cell protein decomposition, both in aerobic and

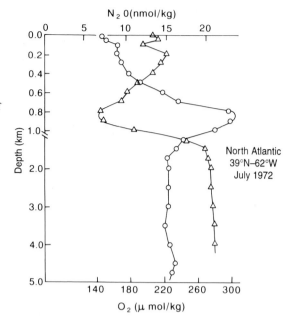

Fig. 4.9. Variation of dissolved oxygen (shown by triangles) and nitrous oxide (N$_2$O; shown by open circles) concentration with depth in the northwestern Atlantic at 39°N, 62°W in July 1972. Note the correspondence of the oxygen minimum and N$_2$O maximum near 1000m. The vertical scale changes at 1000m; note also that the concentration of oxygen is in 10^{-6} moles/kg while N$_2$O is in 10^{-9} moles/kg. Using data from Broecker and Peng (1982).

anaerobic conditions. While it is present in the atmosphere at very low concentrations it has a limited lifetime of a day or so. There seems to be approximate equilibrium for ammonia exchange across the air–sea interface. We saw in §4.1.1 that ammonia can be utilised as a nitrogen source by some phytoplankton, particularly in its dissolved form, as the ammonium ion:

$$NH_3 + H_2O \rightleftharpoons NH_4^+ + OH^- \tag{4.6}$$

Ammonia is a weak greenhouse gas, as well as being a supply of nutrient for marine biological activity.

4.2.3 Sulphureous compounds

Hydrogen sulphide, H$_2$S, and dimethyl sulphide, (CH$_3$)$_2$S, are the two principal sulphureous gases produced by marine biological processes that have climatic influence. Hydrogen sulphide is a major oxidation product of anaerobic decay. H$_2$S can be photochemically oxidised very rapidly to sulphuric acid, H$_2$SO$_4$, in air:

$$H_2S + 2O_2 + \nu \rightleftharpoons H_2SO_4 \tag{4.7}$$

The sulphate particles from this acid can then contribute to cloud condensation nuclei (§3.5.1). However, as H$_2$S is oxidised so rapidly both in air and water (see §3.6) it is unlikely that significant quantities escape from the ocean.

Dimethyl sulphide, or DMS, by contrast, has been observed in considerable concentrations in the marine atmosphere during plankton blooms (2.5×10^{-7} gm^{-3}). It is excreted by plankton during oxidation of its precursor, dimethylsulphoniopropionate, DMSP ($CH_3C_2H_4CO_2SCH_3$). DMS is destroyed within the ocean by several mechanisms. These include biological consumption, biologically and photochemically aided oxidation eventually leading to dissolved sulphate ions, and adsorption onto particles. However, sufficient gas survives these processes to allow transfer of up to a few tens of micromoles of DMS per square metre per day to the atmosphere.

Within the atmosphere DMS can be oxidised via reaction with hydroxyl ions to form sulphur dioxide or methane sulphonoic acid, MSA (CH_3SO_3H). Both sulphate, the product of oxidation of SO_2 (see reaction (3.11)), and MSA form particles that can act as sub-micron cloud condensation nuclei (§3.5.1). The oxidation pathways are

$$(CH_3)_2S + 2OH^- \rightleftharpoons SO_2 + 2CH_4 \tag{4.8a}$$

or

$$2(CH_3)_2S + O_2 + 2HO_2 \rightleftharpoons 2CH_3SO_3H + 2CH_3 \tag{4.8b}$$

Pathway (4.8a) is the predominant one – 80% of DMS is oxidised this way – except in air with low concentrations of NO_x species. In this case pathway (4.8b) dominates. The precise distribution of aerosols derived from SO_2 and MSA for a particular air mass is therefore poorly known. The differing properties of aerosols derived from these two sources makes for difficult assessment of the impact of DMS on climate (§4.4).

DMS release seems to be associated with cell destruction. Oceanic concentrations are greatest not at the height of a spring bloom, when the plankton population is at its highest, but shortly thereafter, when the zooplankton grazing begins to dominate the primary production cycle. The potential importance of DMS for climate means that full discussion of this biological product is reserved for §4.4.

4.2.4 Methyl compounds

Methyl iodide, CH_3I, is another product of algal cell destruction. There is a net flux to the atmosphere, but the main climatic significance of this gas stems from its reaction with dissolved chloride ions to form methyl chloride, CH_3Cl:

$$CH_3I + Cl^- \rightleftharpoons CH_3Cl + I^-. \tag{4.9}$$

The ocean appears to be the most prolific source for methyl chloride, with biomass burning also important. This gas is a natural source of chlorine, the element of concern in the decay of the ozone layer in the stratosphere (§7.2.1).

4.3 Bio-geochemical cycles

All the natural elements cycle between the atmosphere, ocean, biosphere, and geosphere (§1.3.1). A schematic illustration of possible links within a

Fig. 4.10. Schematic illustration of a typical bio-geochemical cycle. Each main component consists of several sub-components, within which cycling can occur, as well as that between each main component. The dotted links indicate direct volcanic input from deep beneath the Earth's surface. Some of the other links are drawn uni-directional; this means only that the major exchange is almost invariably in that direction and not that it is the only route. Boxes with some shading are those in which significant biological activity can occur.

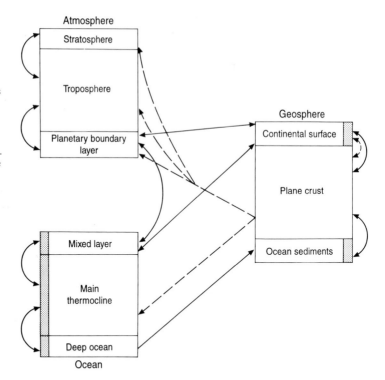

typical bio-geochemical cycle is shown in Fig. 4.10. There are exchanges between the different reservoirs, whose natural equilibrium may be to be in balance, or strongly for exchange in one direction or the other. In addition there may be exchanges within a reservoir, as is particularly true of the ocean. The ocean plays an important role in these cycles through biological processing, and its deposition of sediments (and thus elements) to the under-lying geosphere. Several of the major cycles vital for the existence of life on Earth strongly involve the ocean and its ecology. A full discussion of these cycles is beyond the scope of this book but the following brief overview of the role of marine organisms in several of these cycles summarises the interactions that have appeared in several other parts of the book.

4.3.1   The carbon cycle

The greenhouse gas carbon dioxide is a component of the carbon cycle, the mechanism by which carbon moves between the various chemical reservoirs of the Earth. This was described in §3.3, focussing in detail on the chemical exchange between the ocean and atmosphere, and is shown schematically in Fig. 3.4 (this can be compared with Fig. 4.10). Marine organisms play a small but important part in this cycle. Phytoplankton fix carbon from dissolved carbon dioxide through photosynthesis (reaction 4.1). When these organisms, and the larger species which use them as a food source, die, their remains sink to the sea floor, adding to the sediment. This falling material is known as detritus. The cascade of detritus transfers carbon from the surface waters to the geosphere reservoir. It is estimated that without this natural sink of carbon the equilibrium carbon dioxide concentration before the Industrial Revolution would have been 450 ppm rather than 270 ppm.

There are several complications to this simple picture. Not all the detritus sinks to the bottom of the ocean. A large proportion is re-cycled within the ocean, either through direct consumption by other organisms, or by mixing of the upper ocean in the mixed layer and use of part of the detritus, after bacterial decay, in new production. This re-use is modified by seasonal changes in the mixed layer depth. In winter the mixed layer is deepened by wind mixing and thermal reduction of the upper ocean stratification. This deepening entrains water rich in nutrients, and carbon. In addition the fall-out of carbon from the mixed layer does not provide an escalating sink of carbon dioxide because the biological activity is controlled by a number of physical and chemical properties of the ocean. These have been discussed in §4.1. In the absence of changes to these properties the detritus sink cannot contribute to reducing the impact of anthropogenic increase in atmospheric carbon dioxide. In §7.2.4, however, we will see how there will be feedbacks between physical climatic changes expected from greenhouse warming, some of which are important in controlling marine biological activity.

4.3.2 *The nitrogen cycle*

Micro-organisms control the oceanic component of the nitrogen cycle to a greater extent than for any other geochemical cycle. In coastal regions a large part of the annual budget of nitrogen comes from the land, through run-off. In the deep ocean, however, most nitrogen is cycled within the upper layers, as production is driven by the release of ammonia from dead algae. Some biologically useable nitrogen is mixed upwards from deeper in the ocean, and a small part is added by rainfall from the atmosphere, but these are both relatively small contributions. The most abundant form of nitrogen is the gas, N_2, but most cycling involving marine organisms is between organic and inorganic forms of the element. The global total of coastal and deep oceanic vertical fluxes may be of similar total magnitude despite the smaller (10%) surface area of the coastal ocean.

4.3.3 *The phosphorus cycle*

The chemical composition of organic soft tissue is relatively constant with the ratio of phosphorus:nitrogen:carbon being, on average, 1:15:105. This ratio is known as the Redfield ratio, in honour of the scientist who first demonstrated the ratio's approximate invariance. The ratio of phosphorus to nitrogen in sea water also obeys this ratio in what is thought to be a biologically driven balance. Even in the deep ocean, well away from active microbiological consumption, this ratio holds. Recent research suggests that the C:N ratio within the Redfied ratio may not be as reliable as previously thought. Evidence has been presented of carbon consumption, relative to nitrogen, in both coastal and deep-sea waters significantly in excess (approximately double) of that predicted by the Redfield ratio. This means that current estimates of oceanic uptake of carbon dioxide may be too low. This would have significant consequences for the climatic feedbacks discussed in §7.2.4 and the speed of the climate's response to anthropogenic emissions of CO_2.

The phosphorus cycle is similar to the nitrogen cycle in that much of the exchange within the ocean involves re-processing of material, rather than input from, and output to, outside the system. A small amount of phosphorus enters the sea from land, assisting the rich biological productivity of the coastal waters. Very little enters from the atmosphere. Some is lost to marine sediments, but more than 99.9% of the cycling is between the marine biosphere and the ocean waters.

4.3.4   *The oxygen cycle*

The ocean acts as a sink for oxygen because of its use in respiration by marine organisms. It is also a source through photosynthesis and the release of oxygen during the chemical changes associated with the deposition of marine sediments. Much of the oxygen is re-cycled but the sediment deposition release represents a small leak to the atmosphere which it is estimated would double the atmospheric oxygen in four million years. The excess oxygen is used in the atmosphere during weathering of surface materials to maintain the balance in oxygen levels.

4.3.5   *The sulphur cycle*

Sulphur is a necessary trace element for biological activity. It is provided to marine organisms through re-cycling of dead organic material, dissolution of atmospheric sulphur dioxide, input of sulphates from rivers and precipitation, and anaerobic decay of organic material (see Fig. 4.11 in §4.4). Much of the net sulphur that enters the ocean each year is deposited in sediments. The remainder that is lost from the ocean leaves as biologically produced sulphureous gases, such as DMS. The potential climatic importance of the sulphur cycle, through DMS, merits a separate discussion, given in the next section, §4.4.

4.4   DMS and climate

Dimethyl sulphide, or DMS, was first observed to be present in the ocean in considerable quantity in the early 1970s. Since the mid 1980s there has been considerable interest in this gas as a major source for oceanic sulphate aerosols, which are now thought to form the majority of sub-micron particles in the troposphere. It has even been proposed that climatic feedbacks between algal production, sulphate aerosol levels, and cloud albedo may have exerted a strong climatic control in the past, and be contributing to an amelioration of global warming due to enhanced atmospheric concentrations of greenhouse gases.

DMS is of biological origin, as noted in §4.2.4. It is an oxidation or

Table 4.1. *Gaseous sulphur emission rates (after Schwartz, 1988, units are Tg S $yr^{/1}$)*

Emission Source	Northern Hemisphere	Southern Hemisphere	Global
Marine DMS	17	23	40
Marine H_2S, etc.	5	5	10
Terrestrial biogenic	32	16	48
Anthropgenic sulphur	98	6	104
Total	152	50	202

breakdown product of dimethylsulphonium propionate (DMSP). This latter substance is thought to assist in the limitation of *osmotic* loss of algal cell material to sea water, although the precise purpose of DMSP, and its transformation to DMS, is unknown. DMS is also produced by terrestrial plants. Its impact on the atmospheric aerosol load is less over land, however, as there are many additional terrestrial sources of sulphur.

Table 4.1 shows estimates of the size of the global sources for sulphur emission. Marine DMS contributes about 20% of global emissions (40% of pre-industrial emissions, however), but 80% of those with a marine origin. Long range tropospheric transport of sulphate, the eventual oxidation product of sulphur gases (see reactions (3.11) and (4.8)), is limited because of the hygroscopic nature of the particles and their consequent active participation in cloud, and rain, formation. Measurements over mid-oceanic sites in both the North and South Pacific Oceans suggest that the massive Northern Hemisphere anthropogenic input of sulphur to the atmosphere has only a limited impact on remote marine tropospheric sulphate concentrations, with most of the sulphur being deposited locally or regionally. While emission ratios imply that they should be more like three times Southern Hemisphere values, North Pacific atmospheric levels of sulphate are often substantially less than double those in the South Pacific. DMS may therefore provide about 80% of sub-micron marine sulphate in the Southern Hemisphere, and about 50% of that in the Northern Hemisphere.

There are, however, some significant uncertainties in these figures. The production of sea salt aerosols by breaking waves, discussed in §2.8.2 and §3.5.2, will add sulphate particles to the atmosphere, as well as sodium chloride. The mass proportion of sea salt derived sulphate can be high in some circumstances, but the particles will generally be Giant nuclei, because of their mode of formation, and therefore much fewer in number. The sub-micron sulphate we are considering is often called non-sea-salt, or nss, sulphate because of its distinctly different origin. The quantity of nss sulphate is derived by measuring the sodium or chloride ion concentration in the aerosols and deducting the sea water sulphate:sodium/chloride ratio (see Table 1.4) of the sodium/chloride concentration from the total sulphate level. In addition, it has been found that in the marine boundary layer significant enhancement of the number of larger (super-micron) nss sulphate aerosols occurs. These may be produced by oxidation by ozone

Fig. 4.11. Schematic
illustration of the sources
and sinks of DMS in the
marine boundary layer of
the atmosphere and the
oceanic mixed layer.

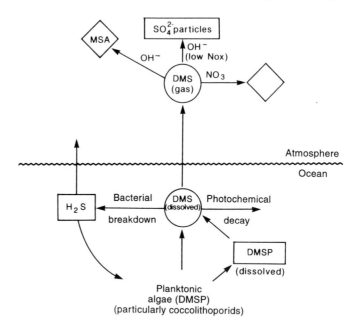

Fig. 4.11. Schematic illustration of the sources and sinks of DMS in the marine boundary layer of the atmosphere and the oceanic mixed layer.

within the very moist atmosphere of this near sea zone. These are then believed to be lost by deposition back into the ocean, thus diminishing the potential for DMS to contribute to mid-tropospheric cloud condensation nuclei. A schematic of the sources and sinks of DMS within the oceanic and atmospheric boundary layers is shown in Fig. 4.11.

With the above reservations, primary production in the ocean is thus responsible for a considerable proportion of the tropospheric aerosols, and thus cloud droplets, that contribute to the climate system in various ways (see §3.5). We have seen that primary production is highly variable in both space and time (§4.1.2); an additional complication with DMS production is its strong dependence on species. The reasons for this, and a good knowledge of the emission rates of different plankton, are presently elusive. The causes are presumably linked to the osmotic processes across cell boundaries, and hence may partially depend on salinity and temperature. The differences in species and seasonal cycles is sufficiently large that the weak productivity of nutrient-limited, or eutrophic, tropical waters produces a similar emission (about 2.2×10^{-3} molm^{-2}yr^{-1}) to more productive temperate localities. Upwelling regions produce slightly more DMS, and the highly productive coastal zones several times as much ($5-6 \times 10^{-3}$ molm^{-2}yr^{-1}). These latter regions, because of their small surface area relative to the global ocean, will contribute rather little to the net global flux of 40×10^{12} gyr^{-1}.

Not all the DMS released by cell decay of algae escapes into the atmosphere. Some is photolysed within the sea (see §3.6). A large proportion is absorbed by bacteria, and oxidised to allow the sulphur to be made available to these organisms. A by-product of this oxidisation is hydrogen sulphide, H_2S. Recent observations in tropical waters suggest that this may be a significant (30–90%) sink for DMS, allowing much less to escape into

Fig. 4.12. Diagram of the feedback loop involving climate and planktonic production of DMS.

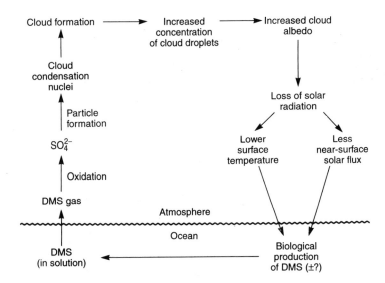

the atmosphere than suggested in the preceding paragraph. Of course, it is possible that the production of H_2S, and its consequent escape to the atmosphere, may partially off-set the atmospheric implications of this bacterial DMS sink. The activity of bacteria in extra-tropical environments may also differ radically from the tropics. Nonetheless, uncertainities about the size of the DMS flux to the atmosphere make present estimates of its climatic significance problematical.

What is this climatic role for DMS, assuming sufficiently large fluxes to the atmosphere to account for the majority of nss sulphate aerosols? Taking a mass-weighted average radius of a cloud droplet to be r, then for a given cloud liquid water amount, L, and droplet number concentration, N, volume arguments show that

$$L = (4/3)\pi r^3 \rho N \qquad (4.10)$$

where ρ is the density of water. Thus, if there is a given amount of liquid water in a cloud a change in number of aerosols, and thus droplets, leads to an inverse change in radius. More droplets lead to tinier droplets. A greater number of such droplets tends to increase the net surface area of the droplets, and hence the cloud albedo by reflecting more solar radiation. Thus if DMS was to increase, the net effect might be a decrease in the input of energy to the climate system, and a consequent global cooling.

This mechanism could be part of a feedback process, illustrated in Fig. 4.12. If climatic warming led to greater oceanic productivity, more nss sulphate aerosols would be produced, leading to more reflective clouds and global cooling. The cloudiness reduces light, and with the cooling, lowers productivity, hence reducing nss sulphate aerosols and allowing the planet to warm again. The climate would be in a state of constant planktonic-aerosol adjustment.

Such a feedback mechanism has an appealing simplicity. However, within the climate system any one feedback loop is only part of a much more complex whole. For instance, the processes linking changing temperatures and planktonic population size and distribution are probably not well understood because we have never consciously observed such a link. In Chapter 7 we will investigate in some detail the various interacting components that may contribute to climatic change over the next century or two.

Further reading

A complete reference list is available at the end of the book but the following is a selection of the best books or articles to follow up particular topics within this chapter. Full details of each reference are to be found in the Bibliography.

Broecker and Peng (1982): An invaluable guide to ocean chemistry. Well written with a very comprehensive list of pre–1982 references. Discusses biological processes where appropriate in text.

Manahan (1990): A comprehensive guide to environmental chemistry with substantial sections on aquatic and atmospheric chemistry with considerable biological discussion. More advanced reading.

Mann and Lazier (1991): An excellent synthesis of marine biology and its interaction with the physical environment.

Parsons *et al.* (1984): A thorough and readable account of oceanic biological processes.

Pinet (1992): A general oceanography text with good introductory material on oceanic biology and chemistry.

Raiswell *et al.* (1980): A good introductory to medium level text on environmental chemistry.

5 *Large-scale air–sea interaction*

This chapter examines two important climatic phenomena in which large-scale interaction between the ocean and atmosphere is a major ingredient. These are: the longitudinal biases in the positioning of tropospheric pressure systems, including associated *cyclogenesis*; and the El Niño/Southern Oscillation phenomenon. Both of these aspects of the climate occur on a permanent, or quasi-periodic, basis and involve strong air–sea physical coupling. Physical, chemical and biological interactions involving longer time scales will be explored in Chapter 6.

5.1 Tropospheric pressure systems and the ocean

In the zonal mean the troposphere has three main meridional circulation cells between the equator and the poles (§1.2 and Fig. 1.5). The surface pressure distribution has far more complexity than this simple model suggests, however. When scanned along a line of latitude, say 35°N, the surface pressure shows considerable variation (Fig. 5.1). In summer the oceanic longitudes show significantly higher pressure than over the continents, by up to 30 mb. In winter some continental areas exhibit higher pressures than over the oceans. This variability is due to heating and cooling affecting the continental air density. The zonal mean pressure, nonetheless, remains higher in the sub-tropics than in regions either poleward or equatorward, in both seasons (Fig. 5.2).

The location of the higher pressure over the oceanic longitudes in Fig. 5.1 is geographically tied to the eastern side of the two ocean basins. This will introduce asymmetry into the forcing of the oceanic circulation by the surface wind field. Such asymmetry is found in Fig. 1.15, but in the west; §2.10.1 showed that the strong, narrow western boundary currents are predominantly caused by conservation of vorticity. These boundary currents bring warm water to sub-polar latitudes (Fig. 5.3) and the resulting latent heat transfers lead to vigorous regions of cyclogenesis in the mid-latitude atmosphere. This tends to bias the sub-polar low pressure formed at the convergence region of cold polar air and warmer mid-latitude air (see Fig. 1.5) to the western side of ocean basins.

These physical links between the oceanic and atmospheric circulation are pursued in general, and for specific ocean basins in the next section (§5.1.1). This then leads to a consideration of the distribution of maritime climate regimes (§5.1.2), and to connections between oceanic variability and

Fig. 5.1. Mean sea level
pressure along 35°N during
Northern Hemisphere
summer (solid line) and
winter (dashed line). The
longitudinal extent of the
main ocean basins is
shown; note that the
Pacific extends on either
side of the dateline
(±180°).

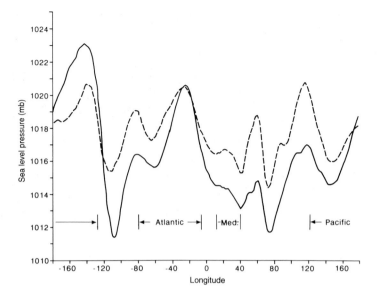

Fig. 5.2. Zonal average sea
level pressure in Northern
Hemisphere summer
(dashed line) and winter
(solid line).

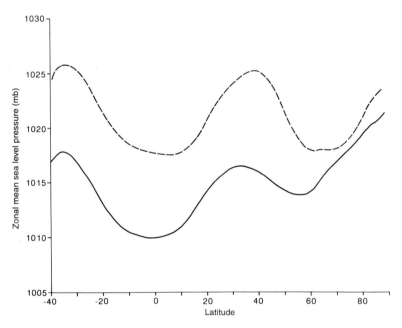

seasonal and inter-annual variation in the climate (§5.1.3). Oceanic impact
on extra-tropical cyclogenesis follows from these investigations (§5.1.4).

5.1.1   The physics of large-scale extra-tropical interaction

The sub-tropical zone is basically a region of atmospheric subsidence (Fig.
1.5). Air converges on this zone at upper levels in the troposphere, leading to
high surface pressure. The descending air is warmed by adiabatic compression, thus lowering the relative density of this air at any given level. This

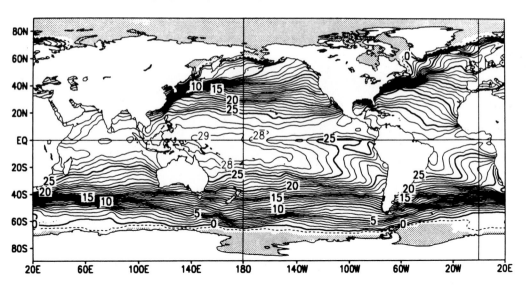

Fig. 5.3. Mean monthly sea
surface temperature for
January 1994. The contour
interval is 1°C, and
negative contours are
dashed. The stippled
regions are ice-covered.
[From the *Climate
Diagnostics Bulletin* (1994a).
Reproduced with
permission of the Climate
Analysis Center, USA.]

serves to partially counter the rise in pressure due to upper level convergence. The descending air also acts to damp any surface-induced atmospheric instability, leading to relatively limited cloud growth.

The actual surface pressure distribution is also influenced by this surface
heating or cooling. Oceanic areas experience a much smaller annual range
in surface temperature than continental land masses (Fig. 5.4) due to the
high heat capacity of water (§1.3). Sub-tropical oceans therefore maintain
relatively consistent pressure patterns over the year, compared to the high,
thermally-driven, variability over land (Fig. 5.1).

The modification of the air temperature by radiation and conduction
from the underlying water also affects the pressure. Warmer water lowers
the overlying pressure, while cooler water raises it. On the eastern side of
sub-tropical ocean basins the atmospheric circulation is equatorward,
because of the geostrophically-driven anticyclonic motion around the
oceanic high pressure regions (see Fig. 1.6). This atmospheric motion is
associated with coastal upwelling in the ocean (§2.10.2). The water on the
eastern perimeter of sub-tropical oceans is therefore cooled, in comparison
with the interior. We have already noted that the western sides of
sub-tropical ocean basins support warm poleward flowing boundary
currents. Therefore a pressure gradient would be expected from east to west
across a sub-tropical ocean. This is evident in Fig. 5.1, and also in Fig. 1.7.
Fig. 5.5 displays the northern summer global surface pressure field showing
the same characteristics. Taking the summer zonal sea surface temperature
gradient across the Atlantic and Pacific Oceans at 35°N, from Fig. 5.3 as
5°C and 9°C respectively, we would expect the pressure gradient across the
North Pacific to be roughly double that across the Atlantic, as is found in
Figs. 5.1 and 5.5.

Over the sub-polar oceans different mechanisms operate. The warm
western boundary currents of the sub-tropical gyres leave the coast and
move into the oceanic interior at the boundary of the sub-tropical and
sub-polar gyres. As these currents move east they spread and cool. Thus the

Fig. 5.4. Annual range of
monthly means in surface
air temperature over the
globe, north of 40°S.
Contours are in intervals of
5°C. Note the strong
damping effect of the
oceans.

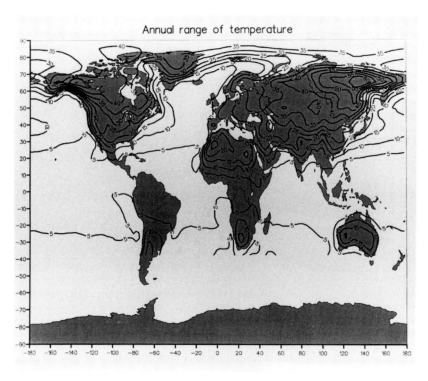

Annual range of temperature

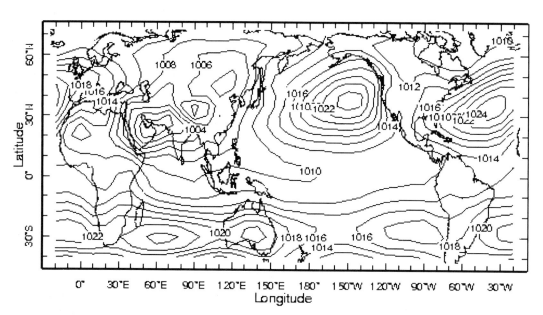

Fig. 5.5. The mean sea level
pressure in July. Contours
are every 2 millibars.

warmest water at these latitudes tends to be towards the west of the ocean basins (Fig. 5.3), and to be associated with lower pressure. This latter tendency is reduced to some extent by the active formation of cyclones – cyclogenesis – in such regions, which often deepen as they move eastwards (§5.1.4).

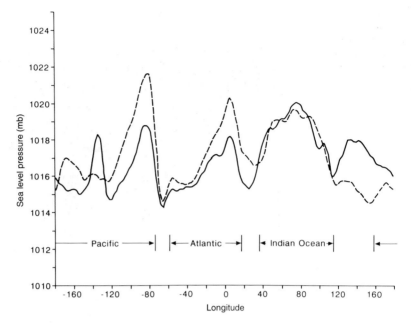

Fig. 5.6. Mean sea level pressure along 35°S during Southern Hemisphere winter (solid line) and summer (dashed line). The longitudinal extent of the main ocean basins is shown; note that the Pacific extends on either side of the dateline (±180°).

These air–sea interactions produce distinct variation in the characteristics of maritime climates around ocean basins. Such properties will be discussed in more detail in §5.1.2. First, however, we need to consider how variation in these interactions from one ocean basin to another introduces significant regional variability.

The Atlantic Ocean sub-tropical high pressure systems are typical examples of the processes discussed above. The longitudinal distribution of pressure at 35°S, shown in Fig. 5.6, correlates well in the South Atlantic sector with that for the appropriate season in the North Atlantic (Fig. 5.1). Sub-polar low pressure in the North Atlantic (the Icelandic Low) exhibits the expected westward bias, allowing for the skewed shape of the basin of the northern North Atlantic. The sub-polar South Atlantic, in common with the South Pacific and South Indian Oceans, however, effectively does not have zonal bounds. A region of active cyclogenesis exists around much of the Antarctic because of the zonal nature of the sub-polar front, and the transfer of latent heat from the relatively warm waters of the Southern Ocean into the cold air off Antarctica (see Fig. 1.6). There are three extraordinarily active regions of southern cyclogenesis: the western Weddell Sea, the region south of the Indian Ocean and the west of the Ross Sea (Fig. 5.7). These tend to be areas of greater heat transfer from ocean to atmosphere.

The ocean–atmosphere interaction of the sub-tropical regions of the Pacific Ocean follows the above model well, despite the large difference in longitudinal extent between the basins in the two hemispheres (Figs 5.1 and 5.6). The sub-polar interaction in the northern Pacific basin is also similar to that of the model, although the oceanic sub-polar gyre is limited in extent by the physical barriers of Alaska and eastern Siberia (Fig. 5.1).

The Indian Ocean and its atmospheric circulation show great contrasts from one hemisphere to the other. The southern Indian Ocean sub-tropical

Fig. 5.7. The density of
cyclogenesis in the
Southern Hemisphere, in
units of numbers of
cyclones ($\times 10^{-4}$) per day
per square degree of
latitude. [From Fig. 5b of
Jones and Simmonds
(1993). Reproduced with
permission of Elsevier from
Climate Dynamics, **9**, pp.
131–45, copyright 1993.]

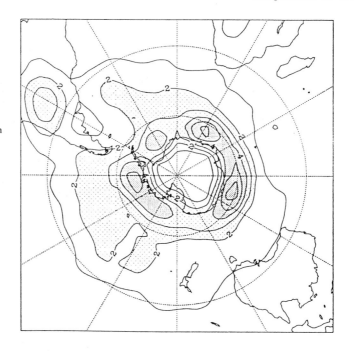

belt shows characteristics of the model (Fig. 5.7). However, the northern basin is subject to complete flow reversal in both the ocean and atmosphere because of the monsoon circulation generated by the Asian landmass. This was discussed in §2.10.4.

5.1.2 *Maritime climates*

Proximity to the ocean modifies the climate of islands and continental boundaries considerably. The sea provides a source of moisture and so promotes cloudiness and precipitation. The limited annual range of sea surface, in comparison to continental, temperatures moderates potential thermal extremes. The prevailing zonal circulation of the lower troposphere, combined with the large-scale interactions between the ocean and the atmosphere discussed in the last section, produces significant geographical variation in the characteristics of these maritime climates.

The anomaly in annual mean surface air temperature with respect to the zonal mean is shown in Fig. 5.8. Contrasts between land and sea dominate this map, but almost as clear are the differences over cool and warm seas. The path of the North Atlantic Drift is clearly seen in the warming of the air over the western North Atlantic extending eastwards, and intensifying, into the Norwegian Sea. Regions of cool eastern boundary currents, and upwelling, lower the atmospheric temperatures of the sub-tropical Atlantic and Pacific Oceans in both hemispheres.

The contrast between warm and cold sea surface temperatures is also apparent in the zonal mean anomalies in cloud cover (Fig. 5.9) and precipitation patterns (Fig. 5.10). These parameters are very sensitive to the air temperature, because of the non-linear nature of the saturation vapour

Fig. 5.8. The anomaly in the annual mean surface air temperature over the globe, relative to the zonal mean. Contours are every 2°C. The annual mean has been evaluated as the average of January and July temperatures.

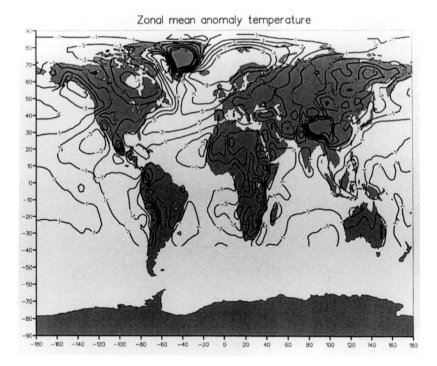

Zonal mean anomaly temperature

pressure curve (Fig. 2.5). The climatic impact of zonal anomalies in temperature are thus accentuated. This is especially so where relatively small temperature anomalies across ocean basins lead to significant changes in evaporation, such as in sub-tropical and tropical regions. For example, the temperature difference from east to west across the South Pacific at 5°S is only +4°C but the cloudiness increases by 40% and the annual precipitation by 1800 mm.

Some coastal regions show strong seasonal variation in climate because of the smaller range in temperature over the sea than the nearby land. The relative tendency for evaporation then alters. For instance, in the southeast sub-tropical Atlantic summer cloud cover is well below the zonal average, because of the cool Namibian Current. However, in the winter the average cloudiness becomes greater than the zonal mean because of the advection of cooled air from southern Africa over the relatively warm sea. Note that the precipitation is still very low as this increased cloud cover is mostly stratus and in, or just above, the marine boundary layer.

The number of factors that contribute to local climate, and particularly marine climate, means that to describe a region's climate in general terms some combination of key variables is necessary. This was one of the aims of classical climatology. It has now fallen from favour to some extent but as the large-scale distribution of *biome* types (Fig. 5.11), and hence surface characteristics of albedo and evapotranspiration, largely follow climate zones it is worth briefly considering the basis of these classification schemes.

Growth of vegetation depends on two basic climatic variables: net radiation and the net balance between precipitation and evaporation.

Fig. 5.9. The anomaly in
the annual mean cloud
cover over the globe,
relative to the zonal mean.
Contours are every 10%.
The annual mean has been
evaluated as the average of
January and July
cloudiness, from recent
satellite data.

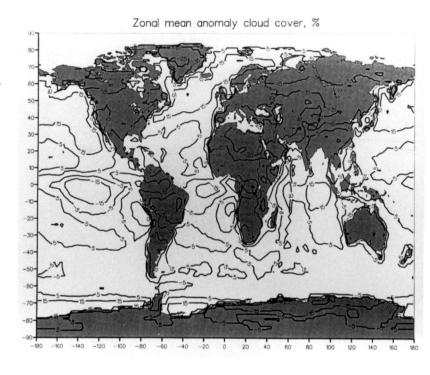

Fig. 5.10. The anomaly in
the annual precipitation
over the globe, relative to
the zonal mean. Contours
are every 300 mm.

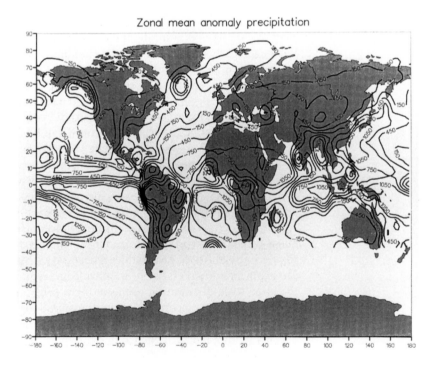

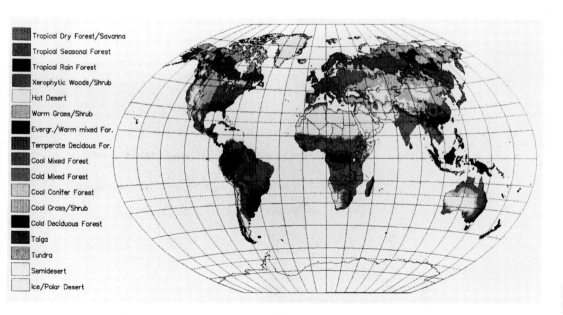

Tropical Dry Forest/Savanna
Tropical Seasonal Forest
Tropical Rain Forest
Xerophytic Woods/Shrub
Hot Desert
Warm Grass/Shrub
Evergr./Warm mixed For.
Temperate Decidous For.
Cool Mixed Forest
Cold Mixed Forest
Cool Conifer Forest
Cool Grass/Shrub
Cold Deciduous Forest
Taiga
Tundra
Semidesert
Ice/Polar Desert

Fig. 5.11. Global biome distributions, based on a model driven by plant physiology, soil properties and climate but in good agreement with mapped distributions. Map drawn from a synthesis of results in Prentice *et al.* (1992).

During the first half of the twentieth century various geographers, most notably Köppen, Thornthwaite, Trewartha and Budyko, produced maps of climatic zones by various parameterisations of these two key variables. One of the simplest (because of the simplicity of the observations required to determine a locality's classification) and most robust is that of Köppen. This relies purely on the relatively simple measurement of temperature and precipitation (but see §6.4.1). The scheme therefore has the limitation of not considering either of the fundamental quantities experienced by plants themselves. Extensive sub-division of this scheme is possible but the principal classifications are defined in Table 5.1 and a global zonation mapped in Fig. 5.12. Comparison with Fig. 5.11 shows a strong correlation between biomes and climatic zones constructed from the Köppen scheme. Possible sub-divisions of the principal classifications are given in Table 5.2.

The geographical province of western Europe and the Mediterranean demonstrates the variation of climatic zone with latitude and continentality (Fig. 5.12). Following the Greenwich meridian from North Africa to the Norwegian Sea one would encounter the latitudinal transitions from sub-tropical semi-arid climate (B) to temperate (C) and finally polar (E). The change from semi-arid to temperate is achieved via a winter wet/summer dry regime (Cs) in which the polar jet only penetrates so far south during the winter. Alternatively, following a latitude line east from Greenwich shows the effect, and extent, of the maritime influence of the North Atlantic in the transition from type C to D (large annual range of temperature).

The spatial extent of these climatic regimes is not fixed in time. During the 70 years from 1871 to 1940, for example, the boundaries between the various sub-types, averaged for 30 year long overlapping periods, fluctuated by 100–200 km. These variations are significantly larger for the maritime/continentality boundary (C/D) than for the more latitudinal

Table 5.1. *Köppen's climate classification*

Classification	Description
A	Tropical rainy climates: temperature of the coldest month $>18°C$
Af	No dry season. At least 60 mm precipitation in driest month
Aw	Distinct dry season. One month with <60 mm precipitation
B	Arid, warm climates: either (i) $R <2T + 28$ if wet summer (ii) $R <2T + 14$ if no wet season (iii) $R <2T$ if wet winter $R =$ annual rainfall(cm); $T =$ annual temperature ($°C$)
BS	Steppe. $R \geq 0.5$ appropriate limit from B
BW	Desert. $R <0.5$ appropriate limit from B
C	Temperate rainy climates: mean temperature of coldest month between $-3°C$ and $18°C$
Cs	Summer dry season. Driest month has $\leq 33\%$ of precipitation of wettest month
Cw	Winter dry season. Wettest month has ≥ 10 times the precipitation of driest month
Cf	No marked dry season
D	Temperate snow climates: mean temperature of the warmest month $>1°C$, coldest month $< -3°C$
Dw	Winter dry season. Wettest month has ≤ 10 times the precipitation of driest month
Df	No marked dry season
E	Polar climates: mean temperature of warmest month $<10°C$

boundaries (a/b/c) (Table 5.2), suggesting climatic variability in the degree of continentality may be greater than latitudinal variation. This variability occurred during a time when the mean global surface air temperature fluctuated by only 0.3°C (§6.4.2). The implications of such movement in reaching potential future climates under the enhanced greenhouse warming hypothesis are substantial and will be revisited in Chapter 7.

The marine extension of the climatic zones in Fig. 5.12 shows the extent of the oceanic regimes discussed earlier in this section. For example, the transition from arid B climates on the eastern sides of sub-tropical oceans to wet A climates on the west occurs at the boundary of the western boundary current. Similarly, the transition from very cold winter D climates on the western side of the northern temperate oceans to the milder C climates occurs along the boundary between the sub-polar gyres and the appropriate western boundary current's oceanic extension. The properties of maritime climates are thus far from being the same everywhere. The position of the continents and the oceanic circulation is a strong moderator of the simple mean picture quoted in the first paragraph of this section. In the next two sections we shall see how these factors influence the atmospheric circulation, and its interannual variability.

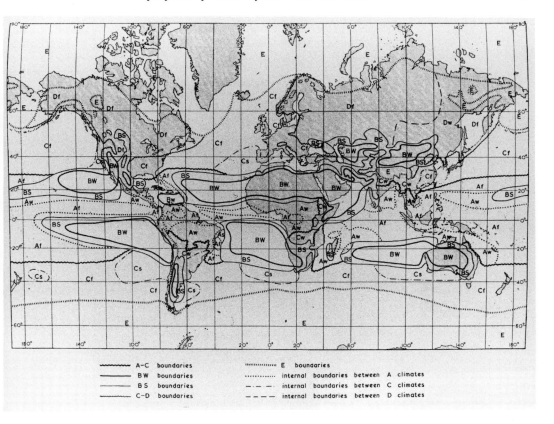

~~ A-C boundaries	ⅲⅲⅲⅲⅲ E boundaries
———— B W boundaries	·········· internal boundaries between A climates
———— B S boundaries	—·—·— internal boundaries between C climates
~~ C-D boundaries	— — — internal boundaries between D climates

Fig. 5.12. Köppen's world classification of climates. [Fig. 11.8 of Lamb (1977), *Climates of the past, present and future, vol. 1.* Reproduced with permission of Metheun and Co.]

5.1.3 *Interannual variability in the atmosphere and ocean*

Correspondence of climatic zone boundaries and major ocean currents immediately suggests strong coupling between interannual variability in the ocean and atmosphere. Movement of the North Atlantic Drift, for example, due to anomalous winds will alter the distribution of latent and sensible heating of the Atlantic troposphere. From resulting alteration to storm tracks downstream this will cause latitudinal movement in the climatic zones of western Europe. Extreme southward relocation of this current, and cooling of the European climate, is probably associated with glaciation, as will be discussed in §6.2.

One of the most significant couplings between the ocean and atmosphere on an interannual timescale occurs in the tropical Pacific – El Niño – and will be discussed in depth in §5.2. This coupling significantly affects the climate of the other tropical ocean basins and, to a lesser extent, of temperate latitudes.

The initiation and movement of extra-tropical cyclones is also strongly linked to the location and intensity of ocean currents (§5.1.4). In this section more general relationships between atmospheric circulation and oceanic

Table 5.2. *Minor sub-divisions of the Köppen climate classification*

Sub-division	Description
a	Hot summer. Warmest month $>22°C$
b	Warm summer. At least 4 months $>10°C$, but $<22°C$
c	Cool summer. Less than 4 months $>10°C$
d	Cold winter. Coldest month $<-38°C$
F	Very cold. Warmest month $<0°C$
g	Ganges sub-type. Warmest month before summer solstice
h	Hot. Yearly mean temperature $>18°C$
H	Polar climate due to high altitude
i	Isothermal. Difference between warmest and coldest month $<5°C$
k	Cool winter. Warmest month $>18°C$ but yearly mean $<18°C$
l	Even warmth. All months between 10°C and 22°C
m	Monsoon climate
n	Foggy but dry
n′	Humid but dry
t′	Warmest in autumn
t″	Coolest just after summer solstice
T	Cool polar (with E). Warmest month $>0°C$
w′	Autumn wet season
w″	Two distinct wet seasons

Fig. 5.13. Precipitation over the Sahel region over the last 100 years, relative to the 1951–80 mean. The data have been normalised with respect to their standard deviation. The spikes show the individual years while the solid curve shows a several year filter to highlight the longer term trends. [Fig. 7.16b of Houghton *et al.* (1990). Reproduced with permission of Cambridge University Press.]

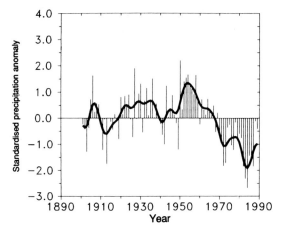

surface conditions will be explored. To illustrate these relationships several specific linkages will be considered: African climate and the tropical Atlantic; movement in the Gulf Stream and western European climate; and western North American climate and variability in the Kuroshio.

Recent decades have seen frequent drought in large areas of sub-Saharan Africa. The most persistent and extreme droughts have occurred in the Sahel, the semi-arid strip south of the Sahara roughly corresponding to the BS zone in Fig. 5.12. Such long-term drought has occurred in the historical past, for example in the early years of this century (Fig. 5.13). Various explanations have been proposed for these long-term changes. These

Fig. 5.14. Sea surface temperature anomalies, averaged over July and August, during the Sahel drought years of 1949, 1970, 1971, 1979, 1981, 1984 and 1985. Drought years coinciding with El Niño events have been excluded. Note the asymmetry about 5°N.

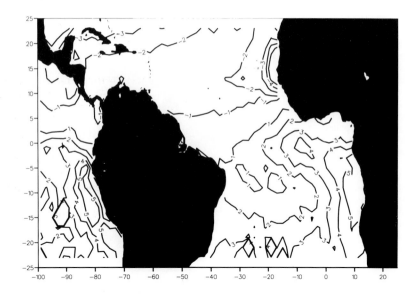

include less northward penetration of the northern summer ITCZ, weakening of the convergence in the ITCZ during this northern excursion, changes in the albedo and evapotranspiration properties of the land surface or changes in the path of the upper tropospheric sub-tropical jet originating over the Indian Ocean.

Teleconnections with strong El Niño events in the Pacific complicate the separation of cause and effect. Droughts occur in such years over more extensive regions of Africa (§5.2.4) than just the Sahel. The drought years unassociated with El Niño events show pronounced anomalies in northern summer sea surface temperatures in the Atlantic. The ocean south of 5°N warms, with respect to the long-term average, while north of 5°N it cools (Fig. 5.14). This anomaly pattern correlates with less northward penetration of the ITCZ into the Sahel during the northern summer. Such a dipolar tendency is also found in the long-term trend in the sea surface temperature field of the tropical Atlantic (Fig. 5.15).

The impact of this anomalous sea surface temperature on the atmosphere is both direct and complex. Changes in temperature alter the latent heat flux correspondingly, by modifying the moisture gradient in the marine boundary layer. The circulation response over the oceans has been to strengthen the sub-tropical high pressure regions over both North and South Atlantic, with the northern zone shifting south and the southern zone shifting east, away from the region of greatest warming. Such reorientation of the source areas for the air carried by the Trade Winds into the ITCZ generally results in a southward shift of the ITCZ trough (Fig. 5.16). Observed intensification of the southwesterly winds in the Gulf of Guinea (Fig. 5.17), which would bring more moisture to the West African area, is compatible with negative correlation between Sahel and Guinea coast rainfall. In the mean this correlation is only weakly significant, in a statistical sense, because of the two ocean–atmosphere states that can be associated with Sahel drought.

Fig. 5.15. The linear trend
in sea surface temperature,
averaged over July and
August, over the period
1946–87. The units are
°C/year. Solid contours
show the zero line or
negative trends (falling
temperature). Note the
extensive warming of the
South Atlantic and the
cooling of the southern half
of the North Atlantic
sub-tropics. The localised
contrasts, such as in the
east Pacific off Peru, are
due to poor data coverage.
See Bigg (1993) for a fuller
discussion.

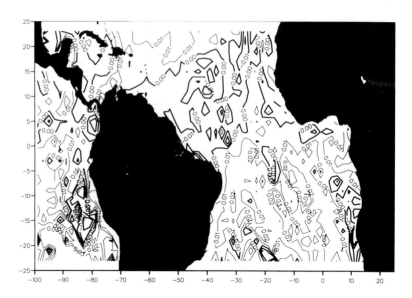

Fig. 5.16. The variation in
the latitude of the ITCZ in
July, averaged over 10°W
to 10°E, between 1960 and
1987. The dashed line
shows the position in
individual years, while the
solid line shows a 5 year
running mean, highlighting
longer term features.

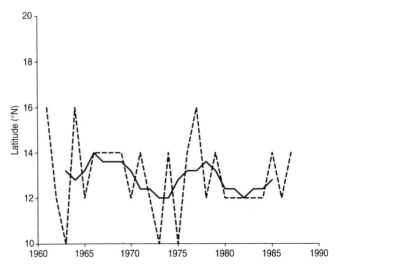

In mid-latitudes there are strong links between sea surface temperatures
and downwind (i.e. eastward) climate. Taking the North Atlantic as an
example, if the North Atlantic Drift is cooled, or displaced southwards, the
climate over western Europe cools and becomes drier because there is less
latent heat flux from the ocean to drive cyclogenesis (§5.1.4). The storm belt
also tends to be displaced southward in this case. In contrast, if the North
Atlantic Drift is warmer than usual more frequent cyclone generation
occurs and western Europe warms, and experiences wetter conditions.

Several examples of these feedbacks are readily found. During 1925–35,
when the globe was warming rapidly (see §6.4.2), boreal autumn sea surface
temperature in the NW Atlantic was frequently anomalously warm. Wetter
conditions than normal were experienced later in the winter over NW
Europe during this time. Similarly, cold periods during the 1690s and 1780s

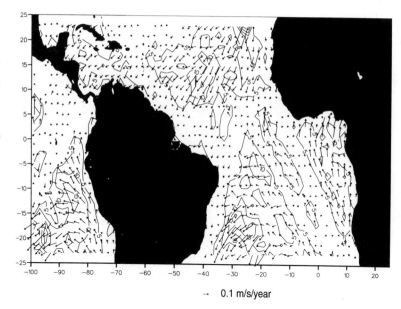

→ 0.1 m/s/year

– part of the Little Ice Age, §6.3.2 – are believed to have been associated with a southerly shift in the Gulf Stream in the NW Atlantic. These ocean–atmosphere links have even been used on a regular basis for making seasonal forecasts for the British climate. They are also found in predictions of the potential future climate under greenhouse warming, as can be seen from comparing the temperature (Fig. 7.17) and precipitation (Fig. 7.19) forecasts for winters under a double present-day CO_2 atmosphere (§7.3.2).

Coupling between the ocean and atmosphere is always occurring, however. This produces striking alteration to the above pattern. For winters in which there is steady cooling of sea surface temperature in the zonal band from 40–50°N the meridional pressure gradient is found to be strong. Anomalously high winds increase transfer of latent and sensible heat from the ocean to atmosphere in such winters, leading to oceanic cooling. Whilst this cooling is taking place the climate of western Europe will tend towards the warm sea surface temperature type of the last paragraph.

The reverse will also occur. Steady warming of sea surface temperatures over winter, relative to their monthly mean, is associated with reduced ocean–atmosphere heat transfer. Combining all these processes gives rise to the opportunity for cycling from one ocean–atmosphere climatic state to another by the modulation of air–sea heat transfer. A schematic of one extreme of this cycle is given in Fig. 5.18; it should be remembered that any regional climatic process does not occur in isolation from the rest of the globe so that the timing, strength and indeed reality of such a cycle in the climate may be difficult to identify. Attaining one end state of this cycle, or oscillation (the North Atlantic Oscillation – NAO), establishes the conditions for the cycle to begin to move towards its opposite state. The warm North Atlantic state pumps latent and sensible heat into the atmosphere, initiating cyclones and strengthening the surface winds. These then take

Fig. 5.18. Idealised relationships between pressure and temperature anomalies associated with the North Atlantic Oscillation. [Fig. 1 of Wallace and Gutzler (1981), *Monthly Weather Review*, **109**, pp. 784–812. Reproduced with kind permission of the American Meteorological Society.]

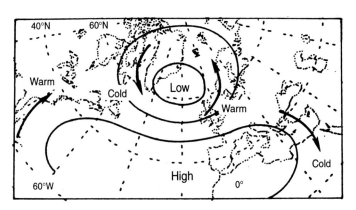

Fig. 5.19. Schematic diagram of the Pacific North American (PNA) pattern of middle and upper tropospheric geopotential height anomalies during a Northern Hemisphere winter, as it coincides with El Niño conditions in the tropical Pacific. The thin arrows indicate a mid-tropospheric streamline (essentially the jet-stream) as distorted by the anomaly pattern. The shaded region indicates where cloudiness and rainfall are enhanced in the generation region. The thick arrows show middle tropospheric winds. [Fig. 11 of Horel and Wallace (1981), *Monthly Weather Review*, **109**, pp. 813–29. Reproduced with kind permission of the American Meteorological Society.]

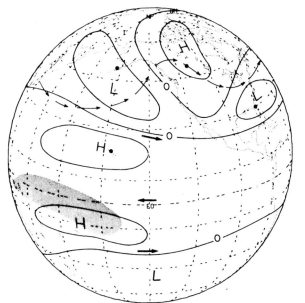

heat out of the ocean, eventually cooling the surface temperatures and pushing the climatic state towards the cold extreme.

Such a mechanism also occurs in the extra-tropical North Pacific. Warm sea surface temperatures over the Pacific west of 180°W influence the atmosphere to produce a wave-like set of mid-tropospheric anomalies – the Pacific North America (PNA) pattern – which propagate across the western hemisphere (Fig. 5.19). The impact on surface climate is to produce a warming of the North Pacific and the continental United States, but a cooling of western Canada and the Canadian Arctic. Heat is again taken out of the ocean, producing increased cyclogenesis and precipitation over mid-latitudes whilst the Canadian Arctic is isolated from the influence of mid-latitude air.

This extreme of the PNA pattern, in generating cyclones, strong winds and evaporation, contains within itself the mechanism by which the central Pacific sea surface warming is reduced. Therefore, if cooling is observed from autumn to spring over the Kuroshio Current in the west Pacific, where

Fig. 5.20. An idealised representation of a mid-latitude depression during the mature stage of its development, following the Norwegian frontal model. The top diagram shows a surface view of the depression, indicating the surface isobaric field by dashed lines and the surface wind direction by continuous arrows (thicker in warm air). The boundary between advancing warm air and cold is denoted by the line with semi-circles (the warm front); the boundary between advancing cold air and the warm sector is shown by a triangulated line (the cold front). The direction of motion follows the jet-stream. A vertical cross-section through the front, AB, is shown below, with typical horizontal and vertical distances. Cloud types are:
Cb – cumulonimbus,
Ac – altocumulus,
Ns – nimbostratus,
As – altostratus,
Cs – cirrostratus.

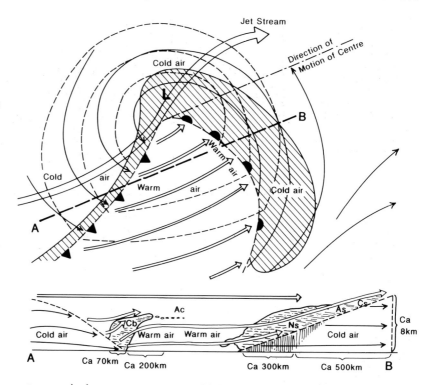

extra-tropical temperatures are highest and cyclone formation more prevalent (§5.1.4), the climate of the Pacific is in transition from one extreme of the PNA pattern to the other. The reverse process, of warming associated with reduced surface winds and cyclogenesis, can also occur. Climatic changes over the North Pacific are, however, not as regular as over the North Atlantic – hence the careful avoidance of the term PNA oscillation – because of the quasi-periodic occurrence of El Niño events, that signficantly influence the climate of at least the Pacific basin (§5.2).

5.1.4 *Oceanic influence on extra-tropical cyclogenesis*

Initiation of cyclones requires energy. A typical low pressure system in the extra-tropics develops at an instability on a boundary between warm and cold air. This can be generated by some combination of factors such as the interaction of the atmospheric flow with mountain ranges – as frequently occurs east of the Rocky Mountains in North America – horizontal shear in wind velocity or local surface heating. These disturbances often develop in a regular way, following the Norwegian frontal model, and are then advected by the mid-tropospheric flow[1]. This is illustrated in Fig. 5.20, with an actual example showing the links between systems in Fig. 5.21; good descriptions of this classic model are found in many texts (see, for example McIlveen (1992)).

Disturbances originating over the ocean such as system A and C of Fig.

[1] The steering level is between 500 and 700 mb. See Appendix D.

Fig. 5.21. A synoptic chart of the North Atlantic and western Europe at 1200 GMT on 5 January, 1994. The isobars are drawn at 5 mb intervals. System B corresponds closest to the idealised frontal system of Fig. 5.20, and has possibly developed due to the influence of the Alps; system C has just reached the occluded phase, where the cold air from the west, travelling faster than the less dense warmer air of the warm sector, has begun to raise the warm air aloft near the centre of the depression. System D is beginning to dissolve, having become almost completely occluded; at A the polar front (separating cold air to the north and warm air to the south) has yet to undergo any distortion, although 24 hours later a new system had begun in this area. [From Bigg (1994).]

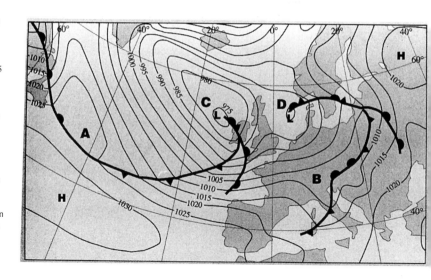

5.21 have an additional source of energy: the latent heat taken from the ocean during evaporation and released in cloud condensation. Frontal zones in the thermal structure of the surface ocean – the Gulf Stream or Circumpolar Current, for example – can provide surface heating discontinuities to initiate or reinforce cyclone development. Such zones can also warm the overlying air in a cyclone, enabling it to hold more moisture and so, through enhanced condensation, become more vigorous.

Overlaying a map of global sea surface temperature with regions of extra-tropical cyclogenesis (Fig. 5.22) highlights the strong links between cyclone generation and the ocean circulation. A majority of cyclones develop over the ocean (although note that a significant minority do not). The most active regions of cyclogenesis are over sharp thermal fronts, associated with vigorous and permanent upper ocean currents: the western North Atlantic (the Gulf Stream), off the east coast of Japan (the Kuroshio), and the southern Indian Ocean (the Circumpolar Current). Coastal regions with strong thermal contrasts can also provide these surface discontinuities, for example along the coast of western Canada and southeast Alaska.

Cyclones are a major means of transferring heat polewards. This is reflected in a general tendency for poleward movement of the vortices once they become distinct features; illustrative tracks over the southern Indian Ocean and the North Pacific, for particular months, are given in Figs 5.23 and 5.24 respectively. If a cyclone's track takes it over a region of warmer ocean, it will tend to deepen and become more vigorous as evaporation is enhanced, adding more energy to the system. The large amounts of energy transferred from the ocean to atmosphere in regions of cyclone generation are seen from comparing Fig. 2.8, the net annual ocean heat balance, with Fig. 5.22. This energy flux can exceed $100 \, \mathrm{Wm}^{-2}$ on an annual basis, and be several times this flux during a typical development.

The development and passage of storm systems over the extra-tropical oceans also has significant impact on the ocean's biosphere. The stirring of

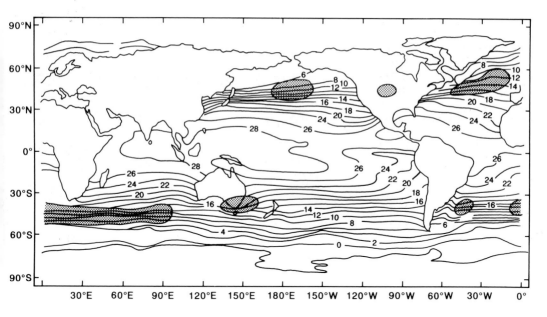

Fig. 5.22. Overlay of regions of cyclogenesis (shown stippled) on the global annual main sea surface temperature distribution. The cyclogenesis regions correspond to areas of high tropospheric eddy kinetic energy.

Fig. 5.23. Tracks of cyclones in January 1958. Only systems that persisted for at least 3 days are included. Black dots show positions 24 hours apart; those systems that formed in the region have their first location ringed. [Fig. 56 of Taljaard and Van Loon (1984) in *World survey of climatology, volume 15.* Reproduced with kind permission of Elsevier.]

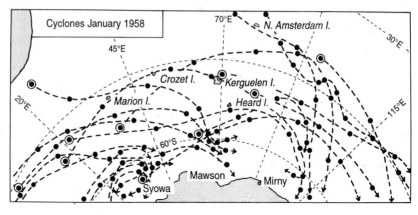

the ocean by strong winds, and particularly strong cooling of the surface ocean during passage of a cyclone's cold sector serves to destabilise the upper ocean. Mixing of surface waters with the more nutrient-rich region below adds vital nutrients to the surface layer. This is most pronounced during the late autumn and winter, but can occur to a more limited extent at other times of the year. The characteristics of the upper ocean before and after the passage of a particular cyclone are shown in Fig. 5.25.

5.2 ENSO: Ocean–atmosphere interaction in the tropics

The most dramatic display of interaction between the ocean and atmosphere on sub-decadal timescales is found in the tropics. This is the quasi-periodic oscillation in climate of the Pacific Basin between 'normal'

Fig. 5.24. Tracks of main
extra-tropical depressions,
and typhoons, in the
Pacific during November,
from the years 1920–40.
[Fig. 15 of Terada and
Hanzawa (1984) in *World
survey of climatology*,
volume 15. Reproduced
with kind permission of
Elsevier.]

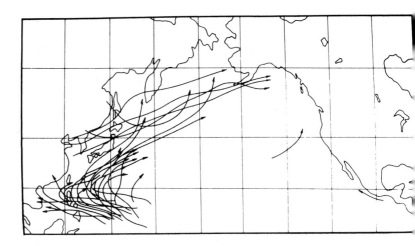

Fig. 5.25. Temperature (°C)
and salinity (psu) profiles,
to 1000m depth, at the
North Atlantic Ocean
Weather Ship Mike 66°N,
2°E, before and after a
strong depression passed
over the station. The
surface layer has been
noticeably deepened,
cooled and freshened by
the passage of the storm.
However, the main
thermocline (600m – 800m)
has also cooled and
freshened, suggesting that
the water column near this
site may have been deeply
mixed by the storm and
the resulting 'new water'
advected under the local
surface-mixed layer.

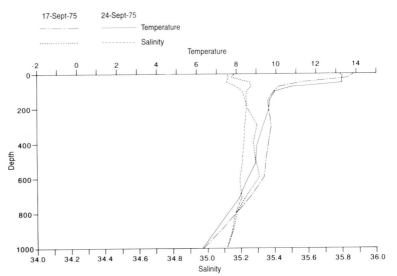

conditions, characterised by strong Trade winds, and 'El Niño' conditions
typified by weak Trades and strong upper ocean warming. This oscillation
has become known as the *Southern Oscillation* and the direct dependence o
El Niño conditions on a particular phase of the Southern Oscillation ha
led to the phrase El Niño Southern Oscillation, or *ENSO*, events becoming
synonomous with the term El Niño. While the Southern Oscillation is a
characteristic of the Pacific Basin it has direct linkages with other region
throughout the tropical climate system and teleconnections with extensiv
areas of the mid-latitudes.

   The first part of our discussion will focus on describing the Souther
Oscillation and El Niño events (§5.2.1). The details of the air–sea interaction
driving the origin and decay of El Niño will then be discussed (§5.2.2). The
driving mechanisms for the Southern Oscillation itself (§5.2.3) are still no
certain but clear links with the Asian Monsoon system exist (§5.2.6). El Niño

events produce an impact beyond the Pacific basin, both in the tropics (§5.2.4) and extra-tropics (§5.2.5).

The extended Southern Oscillation is not the only way in which tropical air–sea interaction occurs, and affects higher latitudes. We have already discussed hurricanes (§2.11.1) and the driving of the tropical ocean circulation by the atmosphere (§2.10.3). There are other processes to consider, such as the influence of tropical maritime air on the mid-latitude climate, through surface or upper air transport, which will comprise the final section (§5.2.7) of the chapter.

5.2.1   *Characteristics of ENSO*

In §2.10.3 and Fig. 2.33 we discussed the basic atmospheric features driving the tropical ocean circulation. Converging, but basically easterly, winds from both hemispheres drive westward currents: the North and South Equatorial Currents. Below the convergence region (the ITCZ), which is usually in the Northern Hemisphere over the Pacific Ocean, an eastwards-flowing counter-current is found. Diverging Ekman transport at the equator leads to upwelling. A strong but narrow equatorial under-current exists to balance the effects of this upwelling, the trans-basin pressure gradient set up by the upward slope on sea level from east to west, and sub-surface convergence on the equator. The atmospheric winds converge to the west of the basin, as well as near the equator. This convergence region is known as the *Indo-Australian Convergence Zone* (IACZ).

This brief description of the 'normal' atmospheric and oceanic circulation in the tropical Pacific depicts one extreme of the Southern Oscillation. The atmospheric circulation can be readily visualised as the surface component of the *Walker Circulation*, illustrated in Fig. 5.26. In the early part of this century Sir Gilbert Walker recognized the potential importance of understanding the Pacific climate for predicting the strength of the Indian summer Monsoon because the Indo-Australian Convergence Zone is the eastern limb of the general convergence over the Monsoon area of the Afro-Eurasian tropics. Consequently, any modulation in the intensity and position of the IACZ is likely to affect the northern summer Monsoon over India.

We shall return to the complex linkage between the Pacific and the Indian Monsoon in a later section (§5.2.6). The circulation pattern proposed by Walker for the Pacific tropical atmosphere, however, is key to understanding interannual climatic variability in the Pacific. In essence, the Walker circulation is analogous to a Hadley cell (see §1.2), but oriented along the equator. The surface winds converge on the low pressure generated by the thermal heating of the warm sea and land surfaces of the west Pacific and Australasia respectively. The converging air is forced to rise, leading to deep and widespread cloud formation, and precipitation. The persistence of this convergence zone is shown by the mean Outgoing Longwave Radiation (OLR) observed from satellites over the Pacific (Fig. 5.27). The value of the OLR reflects the temperature of the emitting surface; low values correspond to thick, high convective cloud while high values

Fig. 5.26. Schematic view
of the Walker circulation
along the equator. Low
level convergence occurs in
regions of convection and
vigorous upward motion.
[Fig. 9.4 of Webster (1983)
in *Large-scale dynamical
processes in the atmosphere.*
Reproduced with kind
permission of Academic
Press.]

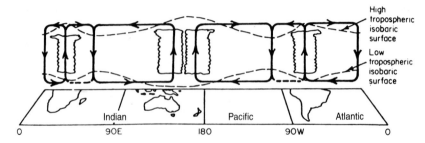

Fig. 5.26. Schematic view of the Walker circulation along the equator. Low level convergence occurs in regions of convection and vigorous upward motion. [Fig. 9.4 of Webster (1983) in *Large-scale dynamical processes in the atmosphere.* Reproduced with kind permission of Academic Press.]

imply no, or generally shallow, cloud. The core of the convergence changes with the season, being centred more south and east in the northern winter than the summer, although a convergence focus for the Pacific is always present. This seasonal modulation is also observed in the intensity of the Pacific Trade Wind circulation: the clearly Pacific-based IACZ of northern winter generates stronger inflow while the more distant IACZ of summer acts as a less efficient sink for lower tropospheric Pacific air. This seasonality is important in understanding the mechanisms for both El Niño initiation and Indian Monsoon variability (§§5.2.2, 5.2.6).

The Walker circulation (Fig. 5.26) is completed by westerly return flow in the upper troposphere (around 200 mb) and sinking of air over the eastern Pacific. The idealised meridional Hadley cell of §1.2 is thus not a feature ubiquitious to all longitudes around the equator, but tends to be restricted to a number of cells of discrete zonal extent. The cores of these Hadley cells tend to be over the sub-tropical oceans, primarily because of heating, and accompanying reduction in surface pressure, over the sub-tropical land masses. The ocean thus assumes another role in determining the climatology of the atmosphere beyond those explored in §5.1.1.

Variation in the intensity of the Walker circulation occurs interannually as well as seasonally. At times the system is strengthened significantly and this affects the climate of North America, if not beyond (§5.2.4). Such periods have become known as La Nina (*the girl*) events because they are the opposite mode of the Southern Oscillation to El Niño (*the boy*; see below for explanation of this name) events. The latter occur when the Walker circulation is significantly weakened. The cycling between these two extremes is known as the Southern Oscillation because of the behaviour in an index that reflects the state of the Walker circulation. This Southern Oscillation Index is the difference in surface pressure, and hence the basic driving pressure gradient force, between the South Pacific High (essentially the subsidence component of the Walker circulation) and the IACZ. Conventionally, the pressure at Tahiti is taken to represent the former and that at Darwin, Australia, to denote the latter.

Variation in this pressure difference since 1934 is shown in Fig. 5.28, and a more detailed examination of the Southern Oscillation Index for the last 20 years is given in Fig. 5.29. In both diagrams the index shows anomalies from the mean monthly difference in the average seasonal cycle, with a 12 month running mean (solid line) fitted to the monthly anomalies. The seasonal cycle is removed because it is not insignificant, as discussed above, and so would unnecessarily complicate the interannual signal. Fig. 5.29

Fig. 5.27. Annual mean out-going long-wave radiation. Contour interval is 10 Wm^{-2}. Values greater than 280 Wm^{-2} (regions of little cloud) are lightly shaded; values smaller than 240 Wm^{-2} (regions commonly having very high cloud) are heavily shaded. [Fig. 2.10a of Hartmann (1994), *Global physical climatology*. Reproduced with kind permission of Academic Press.]

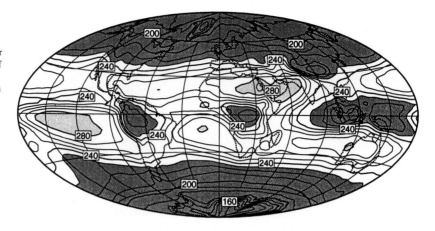

Fig. 5.28. The Southern Oscillation Index (SOI) from 1934 to 1994. The dots indicate monthly averages; the solid line a 12 month running mean. Note that the SOI is relative to the mean monthly difference in sea-level pressure between Tahiti and Darwin, so the seasonal cycle is removed from the basic index. In addition, the data are normalised relative to the data's standard deviation; this gives the scale of the ordinate.

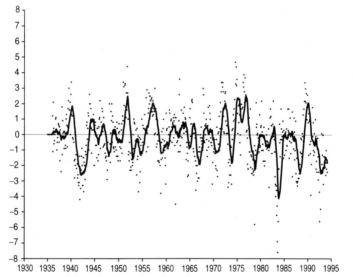

shows the monthly values, as well as this running mean, to illustrate the month-to-month variability. In this diagram the anomalies are shown with respect to the standard deviation of the total series to effectively normalise the data.

These diagrams (Figs 5.28, 5.29) display an oscillation between 'normal' climate, or positive SOI, and ENSO events, or negative SOI. The periodicity has varied over the century, with a greater frequency of ENSO episodes since 1940 than before, but is roughly every 3–4 years. The amplitude of the oscillation is also variable. Some ENSO events, such as those around 1940, 1957, 1982 and 1992, are extremely pronounced and/or prolonged declines in the strength of the Walker circulation. Others denote merely a slackening of this circulation (e.g. 1953, 1963 or 1976). Similarly, there are periods with an enhanced Walker circulation (e.g. 1974 and 1988).

An El Niño, then, is equivalent to the negative SOI phase, or a reduction in the pressure gradient, and thus Trade wind strength, across the Pacific. A reduction in wind strength leads to readjustment in the ocean circulation, in

Fig. 5.29. Detailed SOI
since 1976. The individual
months are shown by
crosses; the solid line is a
five month running mean.
[From the *Climate
Diagnostics Bulletin* (1995).
Reproduced with
permission of the Climate
Analysis Center, USA.]

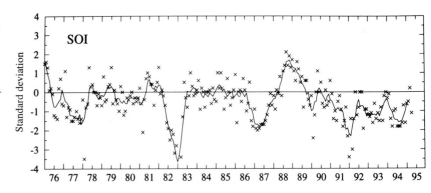

particular a decrease in equatorial upwelling all across the Pacific (see
§5.2.2). Such a decrease in upward transport of cooler water is accompanied
by increased sea surface temperature. From this latter phenomenon is
derived the root of the name of the negative SOI phase: El Niño. Towards
the end of the calendar year the sea surface temperature in the far eastern
Pacific rises rapidly (typically 2–4°C in a month) as the local upwelling is
suppressed. The cause originates with an oceanic response to the seasonal
wandering of the IACZ (§5.2.2). Because this temperature rise occurs near
Christmas the current associated with it is given the Spanish name for the
baby Jesus – The Manchild or 'El Niño'. The onset of a period of
persistently negative SOI often aligns with, and prolongs, the seasonal
Trade wind disruption ultimately responsible for this annual East Pacific
temperature rise. Hence the term El Niño has come to be used for the less
regular but geographically wider and more prolonged, climatic state
associated with a negative SOI.

Each El Niño event is unique, but there are common features which can
be ascribed to the *canonical* event. The surface wind and sea surface
temperature anomalies during such an event, composited from six warm
episodes between 1950 and 1976 by Rasmusson and Carpenter, are shown
in Figs 5.30 and 5.31.

Late in the calendar year prior to an El Niño (Year -1) the South Pacific
High weakens, leading to a decay in the easterly winds in the west central
Pacific. This is accompanied by a rise in sea surface temperature in the west,
due to reduced evaporative cooling. Next, via internal adjustment within
the ocean (§5.2.2), the seasonal warming around the New Year (Year 0) in
the east Pacific is accentuated, resulting in greatly enhanced rainfall along
the South American coast. The warming in the east Pacific is associated
with suppression of the upwelling. Consequent reduction in nutrient supply
disrupts the rich biological productivity of these waters (see Fig. 4.5) and
can drastically affect the Peruvian fishing industry.

These conditions persist for some months, inhibiting the seasonal
cross-equatorial migration of the ITCZ in the eastern Pacific. A further
reduction in the trans-Pacific pressure gradient accompanies these events,
widening the extent of the anomalously weak easterlies, suppressing the
equatorial upwelling and evaporative cooling over a considerable part of
the equatorial Pacific. The IACZ, which is at its easternmost position at this

Fig. 5.30. Surface wind anomalies during a canonical El Niño event. Areas with little data are stippled. Contours are of wind anomalies in ms^{-1}. Panels: (a) August–October of Year -1 (prior to the event); (b) March–May of Year 0 (mature phase); (c) August–October of Year 0 (mature phase); (d) December–February of Year 0/1 (late phase). [Taken from a composite of several diagrams in Rasmusson and Carpenter (1982), *Monthly Weather Review*, **110**, pp. 354–84. Reproduced with kind permission of the American Meteorological Society.]

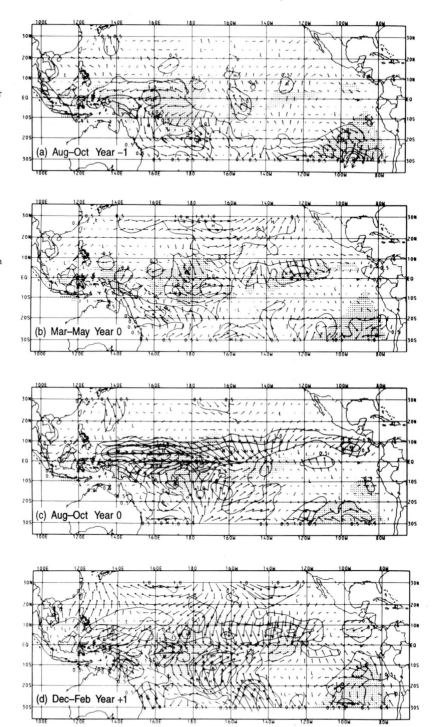

Fig. 5.31. Sea surface
temperature anomalies
during a canonical El Niño
event. Negative anomalies
are dashed; the contour
interval is 0.2°C. Panels: (a)
August–October of Year -1
(prior to the event); (b)
March–May of Year 0
(mature phase); (c)
August–October of Year 0
(mature phase); (d)
December–February of
Year 0/1 (late phase).
[Taken from a composite
of several diagrams in
Rasmusson and Carpenter
(1982), *Monthly Weather
Review*, **110**, pp. 354–84.
Reproduced with kind
permission of the American
Meteorological Society.]

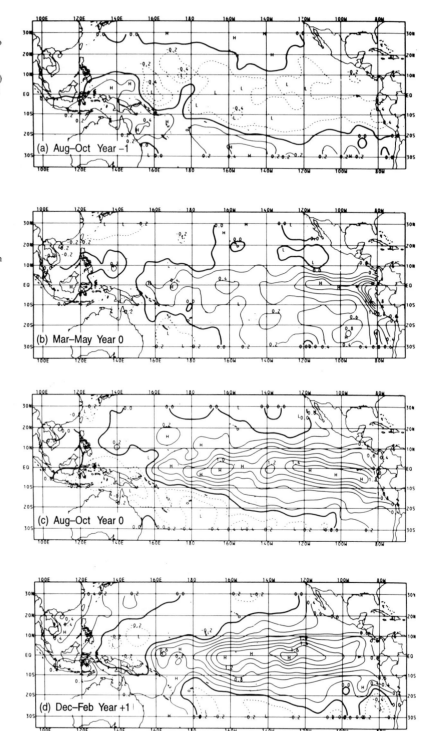

time of year, is drawn further east during the northern spring into the central Pacific, by the warming of the sea surface. These anomalous conditions persist into the following year (Year 1), attaining their maximum extent and severity (the mature phase) at the turn of this year. Anomalous conditions in the east Pacific, however, tend to peak earlier; occasional secondary amplification early in Year 1 can occur, as in 1958, 1973 and 1983.

The end of the event is foretold when cooler surface water appears in the central Pacific in the middle of Year 1 (see §5.2.2). A rapid return to more normal conditions occurs, with the Southern Oscillation sometimes swinging to an extremely positive phase, with strengthened Trades – a La Nina event. These occurred in, for example 1954, 1974 and 1988.

Particular El Niño events demonstrate that the apparent synchronicity with, and amplification of, the seasonal cycle is not essential. The strongest event this century, in 1982/3, began in the middle of the year and followed the conventional timing in the eastward movement of the IACZ, but without an east Pacific precursor. Other events are prolonged beyond the normal life cycle, such as 1939–41 and 1991–4. To explain both the canonical El Niño and these irregular events we turn next to the physics of the air-sea coupling of El Niño.

5.2.2   *ENSO and air–sea coupling*

The mechanism(s) forcing the Southern Oscillation are not well understood. The next section will discuss some of the processes that have been suggested for producing the 3–4 year periodicity. However, the physical processes causing the maturation and decline of the negative, El Niño, phase were unravelled in the 1980s and this section will explore the evolution of El Niño, assuming the existence of the initiating anomaly in the equatorial easterly winds.

Observations of the 1982–3 El Niño were particularly abundant. The strength and widespread impact (§5.2.4) of this event motivated extensive investigation of its mechanics and monitoring of climatic variability in the Pacific (the Tropical Ocean Global Atmosphere Programme – TOGA), resulting in a successful prediction by Cane and co-workers of the succeeding El Niño of 1987, using a relatively simple coupled ocean–atmosphere model. The ultimate finding of this work by many researchers was that El Niño contains the seeds of its own destruction, as will become apparent from the following discussion.

Weakening in the equatorial easterlies was noted in the last section as the initial anomaly in the climate system associated with a developing El Niño. Sudden reduction in the wind stress exerted on the ocean has far-reaching consequences for the upper ocean structure over a wide region. The principal force balances in the equatorial ocean, under mean (or positive SOI) conditions, are shown schematically in Fig. 2.33. When the wind stress underpinning this system is significantly weakened it 'relaxes'. The sea surface slope will diminish, raising sea levels in the east Pacific whilst lowering those in the west. Ekman-driven upwelling will reduce, thus allowing sea surface temperatures to increase. Evaporation will decrease (see equation 2.4), enhancing the local surface temperature increase. Temporarily, the removal of a retarding force to the under-current allows its acceleration.

The ocean needs to adjust over the whole basin to this local anomaly. It does so through the production of internal waves in the upper ocean. These move both east and west from the anomaly, transmitting information on how the ocean needs to adjust. Essentially the ocean mixed layer needs to deepen eastwards, in response to the higher sea level and stronger under-current, while becoming shallower to the west.

The wave transmitted to the east is restricted in latitudinal extent by the Coriolis force, which forces it to decay exponentially, according to the square of the latitude, away from the equator. It is also restricted to purely zonal movement. This wave is known as an internal equatorial Kelvin wave. It has similar properties to *Kelvin waves* that propagate along coasts, and which were first described by Lord Kelvin in the nineteenth century[2]. The speed of propagation of Kelvin waves depends on the density of the water, and the density gradient across the mixed layer, but is about 2.8 ms^{-1}. It thus requires 2–3 months for such a wave to travel across the entire Pacific basin; the rise in sea surface temperature and sea level in the east Pacific therefore lags the onset of the wind anomaly in the west central Pacific by 1–2 months.

Transmission of waves also occurs westward. Like equatorial Kelvin waves, these *equatorial Rossby waves* (see Appendix D) are confined to the equatorial region and tend to decay in a similar manner poleward. However the westward-propagating waves, like mid-latitude Rossby waves, conserve potential vorticity by meridional oscillatory motion. The generating westerly wind anomaly drives a set of these Rossby waves. The fastest travels at roughly half the speed of the Kelvin wave being slowed by the strong vertical shear in the velocity of the basic flow, thus taking 2–4 months to reach the west Pacific coast. These responses to the initial westerly wind anomaly are illustrated in Fig. 5.32.

Several consequences derive from these responses. In the ocean the equatorial Kelvin and Rossby waves alter the thermocline structure towards a less energetic state, even at considerable distance from the initial disturbance. They eventually collide with the coasts bordering the basin and there undergo transformations that, while not immediately affecting the ocean state, are key to understanding the decay of El Niño. The reflection of equatorial Kelvin waves leads to energy being supplied to two further wave processes. Some feeds a direct reflection, which, because of its westward motion, has the properties of equatorial Rossby waves similar to those generated by the initial wind anomaly. The rest excites coastal Kelvin waves along both the North and South American coasts. These decay in amplitude exponentially away from the coast and travel parallel to it.

---

[2] Kelvin waves decay off-shore from a coast in a Gaussian fashion. In the Northern Hemisphere (Southern Hemisphere) they always travel with the coast on the right (left). Equatorial Kelvin waves 'lean' against the equator, as ordinary Kelvin waves 'lean' against the coast. Thus equatorial Kelvin waves are symmetric about the equator.

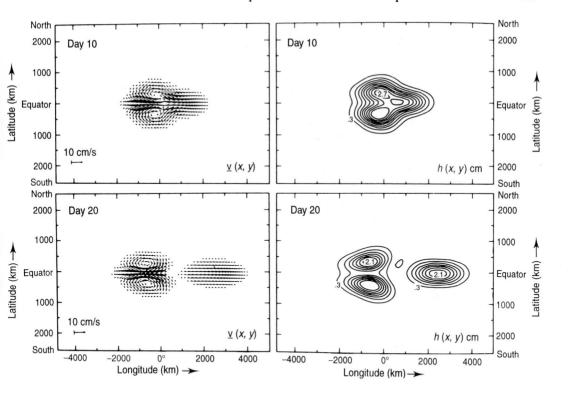

Fig. 5.32. The dispersion of an initial gaussian or bell-shaped thermocline displacement caused by a westerly wind anomaly. The eastward-moving equatorial Kelvin wave and the westward-moving equatorial Rossby waves are clearly seen. The left panel shows the horizontal currents and the right the surface displacement. [Fig. 2 of Philander *et al.* (1984), *J. Atmospheric Sciences*, **41**, pp. 604–13. Reproduced with kind permission of the American Meteorological Society.]

The interaction of Rossby waves with the west Pacific coast is more complex. The boundary is much less well defined, with both large and small islands forming the effective boundary in the New Guinea area. These have coasts that are not oriented north–south, as in the east Pacific. However, it appears that reflection from this broken, irregular coast essentially follows theoretical prediction. Some of the incident energy is lost to heavily damped waves that are restricted to a narrow boundary layer. The rest drives a reflected eastward-propagating, or equatorial Kelvin, wave. The phase of this reflected equatorial Kelvin wave, as for the incident equatorial Rossby wave, is such as to raise the thermocline in the region through which it travels (Fig. 5.33).

The atmosphere also responds to the oceanic changes. The sea surface temperature rise in both the central and east Pacific reinforces the reduction in Trade winds to their west by promoting relative atmospheric convergence into these warmer, less dense areas. This further encourages sea surface temperature rise through reduced evaporation and upwelling, and the eventual eastward shift of the IACZ. The generation of a succession of further equatorial Kelvin and Rossby waves accompanies the continual evolution of the surface wind field. The ocean–atmosphere coupling eventually stabilises during the mature phase of El Niño (Figs 5.30, 5.31), with a stable IACZ in the central Pacific.

The mature El Niño is disrupted by the appearance of cold water in the central Pacific, which reduces the convection, and promotes the re-appearance of the original easterlies. It was earlier stated that the wave

Fig. 5.33. Schematic of internal wave processes stemming from a westerly wind anomaly in the central Pacific. The timescale covered by this diagram is 2–6 months from the initial anomaly.

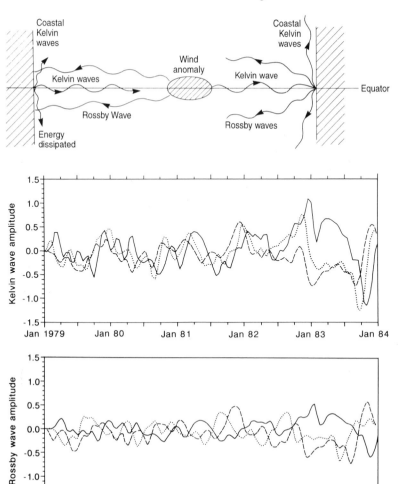

Fig. 5.34. Time series of equatorial Kelvin wave amplitude (upper panel) and the amplitude of the principal equatorial Rossby wave (lower panel) for three regions of the equatorial Pacific. The dashed line shows a west Pacific response (c. 160°E), the dotted line a central Pacific amplitude (c. 170°W), and the solid line shows an east Pacific response (c. 100°W). Equatorial Kelvin wave propagation is shown by a west–east pulse transmission; instances are clearly seen in late 1981, throughout the second half of 1982 (although note that the forcing wind anomaly is here in the central Pacific), and through the second half of 1983 (the Kelvin waves which cause the El Niño event to decay). Equatorial Rossby waves travel east–west at half the speed. These are weaker and harder to see but are visible (from central to west Pacific) in the northern autumn of 1982 and mid 1983. [Adapted from Fig. 6 of Bigg and Blundell (1989), *Quarterly J. Royal Meteorological Society*, **115**, pp. 1039–69. Reproduced with kind permission of the Royal Meteorological Society.]

processes in the ocean provide the clue to this decay phase. The equatorial Kelvin and Rossby wave fields at three locations along the equator during the 1982–3 El Niño are shown in Fig. 5.34. Propogation of equatorial Kelvin waves is visible in the progression of *wave packets* from west to east within time lags of 1–2 months between each location. Equatorial Rossby waves can be seen travelling from east to west with lags of 2–4 months. Remembering that the waves are generated at various sites in the central Pacific, a number of discrete wave events can be seen, with their reflections (equatorial Kelvin waves from Rossby waves and vice versa!). The reflected waves are not always easy to follow, particularly equatorial Kelvin wave reflections from South America, because they are damped by other processes within the ocean. Nonetheless, as 1983 began, the equatorial Kelvin waves generated by Rossby wave reflection from the west Pacific persisted longer. Note that these raise the thermocline. Thus each packet acts to undo the effect of the wind anomaly and eventually raises the thermocline sufficiently for cooler water to upwell and El Niño to abruptly decay (Fig. 5.31).

El Niño initiates its own end. In extreme cases the decay phase coupling pushes the climate into La Nina, as in 1988. Few phenomena better illustrate the complexity of the interaction between ocean and atmosphere.

5.2.3 The ENSO cycle

The periodicity of the ENSO cycle in Fig. 5.28 is on first inspection striking, but closer examination shows that this regularity is illusory. Since the 1950s El Niño events have occurred every 3–4 years, almost without fail. However, there was not a single event between 1932 and 1939; the return period in the early eighteenth century approached 10 years. The duration of events can also be much longer than the 1–2 years of the canonical event. From 1939 to 1944 the SOI was only positive for a short time, in 1942/3. Beginning in late 1990 the SOI remained moderately negative until 1995. This variability on an underlying cycle points to two factors that will need to be considered in trying to understand the ENSO cycle: a mechanism with a several year period, but activated by a random triggering device.

The ENSO cycle may be self-contained, driven by external agency or be purely stochastic. Random triggering may therefore merely be necessary for initiation of El Niño from a pre-conditioned ocean–atmosphere state or act as the determinator of an apparent cycle. However, if the Southern Oscillation is not stochastic then the vacillation must be controlled by a component of the climate with long memory. The atmosphere in isolation mixes thoroughly in a few weeks. Thus, neglecting, for the moment, influences outside the conventional climate system this means that the controlling mechanism must be in the ocean.

Whatever the process of initiation there must be sufficient energy in the flywheel of the tropical Pacific Ocean to fuel the massive reorganisation of the climate system concomitant with El Niño. This energy is stored in the *warm pool* of the western Pacific, beneath the northern winter position of the IACZ. Here surface temperatures can exceed 30°C and mixed layer depth can be 150m. Consistently, prior to El Niño the heat content of this area is above average. There is some correlation between the excess heat stored and the severity of the succeeding event. Prior to 1939, 1957, 1972 and 1982 a number of years passed without El Niño events, or the occurrence of only relatively minor events. These years then heralded severe events. However, the energy is recharged within a year or so after an El Niño (Fig. 5.35), implying that the storage mechanism is not controlling the ENSO cycle.

The secondary importance of the warm pool was further supported by the 1991–4 event. This El Niño had the character of two events, separated by a period of a few months in the middle of 1992 when warm conditions in much of the Pacific were relaxing. The second phase was initiated by strengthening atmospheric convection (and thus the re-appearance of westerly anomalies to its west) that had failed to shift as far west as Indonesia at the close of the first phase. The west Pacific in 1992 had

Fig. 5.35. Upper layer
volume anomaly in the
tropical Pacific, averaged
over 15°N to 15°S. The
units are in $10^{14}m^3$,
relative to its mean value
of ∼ $70 \times 10^{14}m^3$. The thin
line shows monthly values
(retaining the seasonal
cycle); the thick line shows
a 12 month running mean.
[From the *Climate
Diagnostics Bulletin* (1994c).
Reproduced with
permission of the Climate
Analysis Center, USA.]

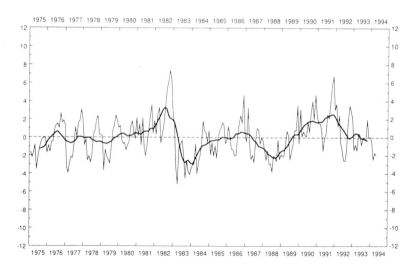

Fig. 5.35. Upper layer volume anomaly in the tropical Pacific, averaged over 15°N to 15°S. The units are in $10^{14}m^3$, relative to its mean value of ∼ $70 \times 10^{14}m^3$. The thin line shows monthly values (retaining the seasonal cycle); the thick line shows a 12 month running mean. [From the *Climate Diagnostics Bulletin* (1994c). Reproduced with permission of the Climate Analysis Center, USA.]

anomalously cool surface waters, which discouraged the return of the convection region. The warm pool was therefore not sustained during the first phase of this event, in contrast to the 1940s event, and so cannot be linked to the 1991–4 event's longevity.

Longer time scales in the ocean can be found in the dynamics of internal waves in mid-latitudes. These Rossby waves (see Appendix D) require about two years to cross the Pacific basin from east to west at sub-tropical latitudes. Theoretical arguments presented by Graham and White in the late 1980s imply that the coastal Kelvin waves directed polewards along the western coast of the Americas as a result of equatorial wave dynamics in an El Niño (see §5.2.2) can initiate the generation of westward-propagating Rossby waves. This must rely on interaction of the coastal waves with irregularities in the shelf topography. Upon reaching the west Pacific coast these could then feed energy into equatorward-propagating coastal Kelvin waves. In turn, at the equator these convert to equatorial Kelvin waves, acting to depress the thermocline and limit equatorial upwelling. Consequent raising of sea surface temperatures in the west central Pacific weakens the Trades and an El Niño begins (Fig. 5.36).

This proposed cycle contains the right inherent time scale but is very idealised. Clearly the mechanism is not invariant, or ENSO would be more regular than observed. Mid-latitude air–sea interaction can weaken, or slow, the Rossby wave propagation. Conversion of energy to coastal Kelvin waves along the west Pacific coast is problematic, at least in the northern Pacific, because of the broken nature of the island barrier. Unrelated enhancement of the equatorial easterlies in the atmosphere may counter the final impact of the cycle. Nonetheless, some observational evidence for such waves has been obtained in North Pacific sea surface temperatures. Numerical modelling of the prolonged negative phase of the SOI in the early 1940s has suggested these waves may have been responsible for the longevity of this event. The different widths of the sub-tropical north and south Pacific led, in this model, to propagation times for coincident Rossby

Fig. 5.36. Schematic of the Graham and White (1988) hypothesis linking one El Niño event to its successor through internal dynamics within the Pacific Ocean. The left and right extremes represent idealised tropical coastlines of the west and east Pacific respectively. An idea of the time scale of the different dynamical components of the cycle are indicated in parentheses.

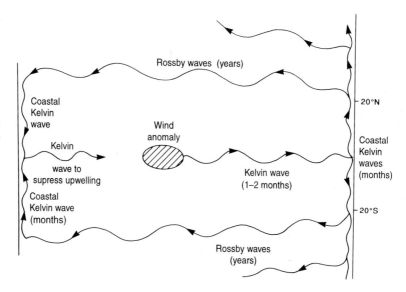

waves differing by almost a year between hemispheres. There is currently no evidence, however, of such processes sustaining the 1991–4 event.

More recently, satellite observations and numerical model simulations of sea surface height over the North Pacific following the 1982/3 ENSO have provided further confirmation of mid-latitude Rossby wave generation. Such waves from the 1982/3 event, travelling poleward of Hawaii at 35°–40°N, and thus at significantly slower speeds than the Rossby waves hypothesised above for the early 1940s, may be responsible for northward displacement of the Kuroshio Current during the early 1990s. This would shift the region of cyclogenesis in the North Pacific and have a downstream impact on climate over western North America. The climatic impact of strong El Niño events may therefore be both concurrent and delayed.

Rossby waves do therefore seem to be important in providing a mechanism for the ENSO cycle, and allowing decadal scale after-effects of strong warm events. However, the difficulty in identifying them observationally suggests that the hypothesis linking Rossby waves to El Niño generation may be part of a more complex instability in the coupled ocean–atmosphere system.

A number of proposals for more remote forcing mechanisms possessing a periodic, or pseudo-periodic, nature for the ENSO cycle have been made. These include the influence of solar variability, perhaps through the *Quasi-Biennial Oscillation* (QBO) of the stratosphere, and volcanism. Association of times of high frequency ENSO cycle with low solar output (few sun-spots), and vice versa, is found for the past 250 years. Relatively few periods of extremes in either variable occurred over this period, however, so the relationship could be coincidental. Solar variability has an impact on climate directly through changing the basic energy supply (see §7.1.1). The limited fluctuations associated with the 11 year cycle in sun-spots are not sufficient to have a significant effect on the climate, although it does modulate the strength of the QBO. This phenomenon is a zonally symmetric oscillation in the circulation of the lower stratosphere between alternating easterly and westerly winds. Each regime is symmetric about a peak amplitude at the equator and the oscillation has a period of 24–30 months (hence quasi-biennial). The circulation in the stratosphere is maintained by vertically propagating waves from the tropical atmosphere, driven by the annual cycle. These are analogous to the equatorially-trapped waves in the ocean discussed above. The QBO is modulated by these waves and may feed back to the tropical atmosphere, affecting the ENSO cycle. Again this link may modulate ENSO but is unlikely to determine the underlying periodicity.

Another external forcing of the climate that has been linked to ENSO is volcanism. Major eruptions can inject considerable quantities of aerosol into the stratosphere. The aerosol of eruptions of tropical origin is more easily mixed over the globe to produce a distinct warming of the stratosphere, and cooling of the troposphere. A number of El Niño events of this century occurred shortly after such eruptions; however, many cannot be associated with such forcing. The thermal inertia of the ocean suggests that during volcanically-driven cooling of the troposphere the land-based convergence zones, such as the IACZ, may be weakened. This may push the Southern Oscillation towards its negative, El Niño, phase. The evidence for this mechanism is purely circumstantial; once again, if it sometimes acts as a trigger it is not associated with all El Niños and so does not determine the basic periodicity of ENSO.

A number of potential trigger mechanisms for El Niño, given a suitably pre-conditioned ocean, have been discussed above. A few others also deserve mention. The generation of a pair of hurricanes across the equator in the west Pacific, which does sometimes occur, would temporarily produce pronounced westerly anomalies in the tropical wind field. The resulting oceanic Kelvin and Rossby waves could then lead to El Niño. A way of weakening the IACZ at some distance from the Pacific is through the winter snow cover over Tibet. Abnormally large winter snow cover results in a weaker summer monsoon over India, because the potential for prolonged warming of the region during summer is reduced (§5.2.4 and §5.2.6). This could weaken the easterlies over the Pacific feeding into the convergence zone, leading to El Niño. Neither of these mechanisms explains all El Niño events, nor do they invariably lead to one.

Note that the combination of all these different triggering mechanisms could induce an ENSO-like cycle purely stochastically. It is necessary for the Pacific Ocean's heat reserves to be recharged, but then any of these mechanisms may, in the right circumstances, act as a trigger for the next El Niño. However, the relative success of simple coupled models of the tropical ocean and atmosphere to predict the recent fluctuation in the ENSO cycle suggests that the origin of the cycle lies in the ocean. Various mechanisms may act as triggers but the ocean is pre-conditioned in a semi-regular fashion, perhaps by some process such as the Graham and White hypothesis or a more complex mode of ocean–atmosphere interaction.

Fig. 5.37. Global impacts of El Niño. [Fig. 4 of Bigg (1990), *Weather*, **45**, pp. 2–8. Reproduced with kind permission of the Royal Meteorological Society.]

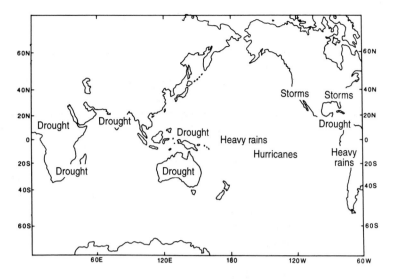

5.2.4   *The impact of ENSO in the tropics beyond the Pacific Basin*

The intimate relationship between the Indian Monsoon and ENSO has been alluded to on several occasions. Both phenomena are tied to the IACZ. The impact of the ENSO cycle extends beyond the Indian Ocean, however, to the entire tropics and substantial areas of the mid-latitudes. In this section we will examine the tropical impacts, leaving the extra-tropics for §5.2.5.

Exceptional weather associated with El Niño events is shown, on a global basis, in Fig. 5.37. The abnormal climate of the tropical Pacific is directly due to the processes discussed in §5.2.2. The eastward movement of the IACZ leads to droughts over Indonesia and Australia, but unusually heavy rains in the normally dry central Pacific. During severe warm events sufficient heat and moisture is added to the central Pacific atmosphere for the generation of hurricanes. Tahiti very rarely experiences hurricanes, however, during the El Niño of 1982–3 five swept the islands. Heavy rains also occur over the west coast of South America, a desert under normal conditions. These are due to atmospheric convection, initiated by the large rise in sea surface temperatures due to the elimination of the local upwelling along this coast.

The disruption of the atmospheric circulation over the tropical Pacific requires adjustment to the mechanics of the entire tropical climate. Displacement of the IACZ into the Pacific results in the establishment of a separate convection zone over southeast Asia during the Monsoon season. This draws air from a much reduced area compared to a normal Monsoon, because of the adjustment of the tropospheric circulation to competition from the central Pacific convergence. Weak Monsoon rainfall, and drought in extreme cases, is the result (Fig. 5.39 in §5.2.6). Sub-normal rainfall during the Indian Monsoon is, however, not invariably coupled with El Niño, as will be shown in §5.2.6.

Failure of seasonal rainfall is El Niño's signature in other tropical regions

Fig. 5.38. Annual rainfall
over NE Brazil, 1912–85.
The annual totals have
been normalised with
respect to the standard
deviation of the series.
Note the tendency for a
bi-modal distribution,
corresponding to whether
the Amazonian convection
region is to the west
(negative anomalies) or the
east (positive anomalies).
[Fig. 5.17a of Goudie
(1992), *Environmental
change.* Reproduced by
permission of Oxford
University Press.]

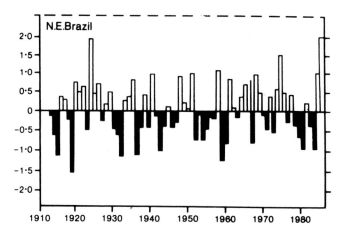

Fig. 5.38. Annual rainfall over NE Brazil, 1912–85. The annual totals have been normalised with respect to the standard deviation of the series. Note the tendency for a bi-modal distribution, corresponding to whether the Amazonian convection region is to the west (negative anomalies) or the east (positive anomalies). [Fig. 5.17a of Goudie (1992), *Environmental change.* Reproduced by permission of Oxford University Press.]

too. In normal years a convergence zone is centred over the heated land mass and coastal seas of northern South America. During the *boreal* winter it is located over the Amazon basin, but moves to Central America and the Caribbean during the summer (see Fig. 5.16). During a warm event the convergence re-orients over the eastern Pacific, causing the annual wet season further east and north to fail.

Drought over much of Africa also coincides with El Niño. The dislocation of the IACZ not only limits the strength of the Indian Monsoon but also the seasonal rainfall in southern Africa that follows the austral summer migration of the Indian Ocean ITCZ (§2.10.4). The ITCZ over the Indian Ocean is weakened during El Niño, because of the changes in the lower troposphere convergence pattern. When the ITCZ moves south in the latter part of the year the intensity and number of convective storms is much reduced. This limits the effect of the austral summer rainy season in sub-tropical southern Africa following an El Niño.

The Sahel rainfall (§5.1.3), a boreal summer feature, fails during strong El Niños. This is largely linked to disruption in the circulation over the Atlantic associated with the westward re-location of the South American convergence zone. Weakening of the Indian Monsoon, and its upper tropospheric easterly jet, may also contribute to the decline in rainfall. The enhanced convection over the east Pacific draws stronger winds from the Atlantic than normally flow into the Central American convergence zone. Stronger winds act to cool the tropical Atlantic, which was noted in §5.1.3 to result in a decrease of the west African summer rainfall.

Following some intense El Niño events the Atlantic can experience similar behaviour. In 1983 the Atlantic Trades strengthened significantly because of the particularly strong convection region over the eastern Pacific during the 1982–3 warm event. Consequently Sahel rainfall was much reduced. In the following year rainfall over the normally dry areas of northeast Brazil and southwest Africa was exceptionally high, while the Trade winds were weaker than normal. Similar behaviour occurred after the severe El Niño of 1972–3. The reasons for this behaviour are not well understood. The tropical Atlantic can experience El Niño-like events independently of the state of the SOI. The Amazonian convergence region is coupled to subsidence over the south east Atlantic in a manner analogous to the Pacific's Walker cell. With much less regularity than the Southern Oscillation (Fig. 5.38) the convection penetrates further east during the austral summer, although remaining over South America. Weakening of the Atlantic Trades leads to less upwelling off southwest Africa and local convection. However, this process is limited in duration, because the convection is drawn across the equator into the northern hemisphere as the northern summer begins. As the convection never moves out to sea, but is restricted to land, which has a much faster thermal response than the sea, the prolongation of the Atlantic warm event beyond the austral summer is not possible.

5.2.5 *The impact of ENSO in the extra-tropics*

ENSO, or at least the negative phase of the SOI, has a global effect on the tropical climate. The large-scale disruption to the convergence zones of the tropics will not only change the tropical subsidence areas, as in the Walker cell, but must also have an impact on the meridional, or Hadley cell, subsidence. Ultimate impact on the mid-latitude storm belts, or even polar latitudes, can be imagined. This process of a climatic shift in one area affecting the climate in a far removed locality is known as teleconnection. The consequences of El Niño discussed in the previous section are teleconnections, but the term is most commonly used when referring to tropical–extra-tropical climatic links.

The clearest ENSO impact on the climate of the extra-tropics is through its resonance with the PNA pattern described in §5.1.3. A schematic of the accentuation of this pattern by a Pacific warm event is shown in Fig. 5.21. The pattern is thought to relate to an instability in the mid-tropospheric flow which steers the cyclone paths (§5.1.4). A simplified view of this instability can be gained by considering the impact on mid-latitude convergence of El Niño's anomalous convection in the central Pacific. We have seen that the high pressure systems in the Pacific sub-tropics weaken during a warm event. The sea beneath will also tend to warm, as the weaker Trades lower evaporative cooling. Both of these effects will act to decrease the subsidence in the Pacific sub-tropics, allowing further penetration equatorward of the north Pacific storm belt than normal (forming a *trough*). The wave so generated in the upper tropospheric flow moves anomalously north over western Canada. Intrusion of warmer air follows, and central Canada and the western United States experience warmer temperatures than normal in the winter following El Niño onset (that is, Year 0/1 in the terminology of §5.2.2). Greater rainfall than normal is associated with this spread of moist tropical air over the western United States.

The tropospheric wave instability is of *wave number* 4. This means that there are roughly 90° of longitude between successive troughs. Consequently, in eastern North America another equatorward excursion of the tropospheric steering flow occurs, bringing cold and wet weather to the Year 0/1 winter in the southeastern United States.

This pattern, although becoming less clear away from the region of

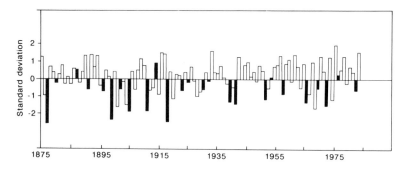

generation of the anomalous circulation, can also be discerned over western Europe. Winter cyclone tracks of Year 0/1 tend to be pushed south over the British Isles, in the third trough of the wave. This tends to bring cooler and wetter conditions to western Europe. The fourth trough should occur over central Siberia, but the winter circulation is strongly modified by the dominant Siberian High pressure and little impact of El Niño on northern Eurasia is observed.

The formation of wave instabilities in the upper troposphere of the Southern Hemisphere is much less common. Depression tracks, however, tend to be further north across the South Pacific and southwest Atlantic generally. Thus the Year 0/1 austral summer in New Zealand and Argentina tends to be wet, particularly in the former where the summer is also anomalously cold.

In the years of extreme positive SOI – La Nina or cold events – the PNA pattern operates in reverse. The effects tend to attenuate to a greater degree downstream of the west Pacific. However, the cold outbreaks over central North America associated with La Nina can be extreme, as in 1988 for example.

5.2.6   *ENSO and the Indian Monsoon*

Monsoons are, to a first approximation, giant sea breeze systems. The land is heated during summer, creating a pressure gradient from the cooler sea towards the land. During the winter the land cools more than the ocean and the pressure gradient is reversed. The Indian Monsoon (§2.10.4) is often quoted as the archetypal monsoon. It has, however, two complicating characteristics.

Changes in the Pacific Basin climate are directly linked by movement of the ITCZ to those over southern Asia (§5.2.4). The direction of impact nevertheless appears to be one way: from the Pacific to Asia. El Niño events are almost invariably associated with weak, or low rainfall, Monsoon years[3]; such Monsoons, however, also occur in other years (Fig. 5.39). This appearance of unilateral effect may, nonetheless, be illusory. Attempts to predict the evolution of the SOI – a *proxy* for the Pacific climate – from boreal winter to summer lack accuracy, whether using statistical or coupled

---

[3] Note that there are exceptions. The long-lived 1991–4 El Niño coincided with a period of normal Monsoon rainfall over the Indian sub-continent.

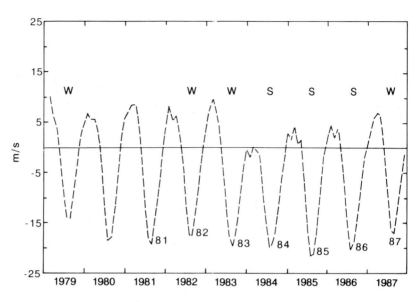

Fig. 5.40. Three-month running average of the 200 mb (upper troposphere) zonal wind over southern Asia (5°–20°N, 40°–110°E) for 1979–87. Positive is a westerly wind. The occurrence of weak (W) and strong (S) Monsoon rainfall over India is indicated. Note that before a strong Monsoon the upper tropospheric winds tend to be weak. [After Webster and Yang (1992), *Quarterly J. Royal Meteorological Society*, **118**, pp. 877–926. Modified with the kind permission of the Royal Meteorological Society.]

numerical modelling techniques. The spring is a period of weak circulation in the Pacific, when random forcing can send the recent climatic tendency askew. Forecasting the strength of a succeeding summer Monsoon over southern Asia seems to have greater potential because there is, at least for recent decades, a good correlation with the upper tropospheric wind field (Fig. 5.40). Weak Monsoons are heralded by significantly westerly winds at heights around 200 mb, at which level the winter flow is usually quiescent.

A further complication arises when considering the precise heating over the southern Asian land mass. The basic monsoon mechanism (§2.10.4) is that heating the land in summer lowers the local pressure and drives an inflow of air from the ocean. However, numerical modelling experiments which flatten the Tibetan plateau – in reality some 4.5–5 km in height – to sea level show very weak monsoon behaviour. Thus the establishment of the pressure gradient depends on a heating differential well into the lower troposphere.

The reasoning can be seen from considering Fig. 5.41. The solar energy incident on southern Asia heats the ground, which in turn heats the air above it through transfer of sensible heat. If this ground is elevated the air above is less dense than if the surface was at sea level. Thus, the same amount of energy will increase the temperature more than at sea level. Therefore summer temperature on the Tibetan plateau would be expected to be greater than nearby areas closer to sea level. A number of other factors will influence local temperatures but comparing the average July temperature, T_7, at Lhasa (29.7°N, 91.1°E, 3659 m, $T_7 = 15.5$°C) with two stations on either side of the Tibetan plateau at a similar latitude, Ya-an (30.0°N, 103.1°E, 637 m, $T_7 = 26.0$°C) and Ludhiana (30.9°N, 75.9°E, 247 m, $T_7 = 31.0$°C), confirms this hypothesis. The average decrease of temperature with height in the atmosphere, due to expansional cooling, is about 6°C per 1000m. Lhasa is thus 5–8°C warmer in the summer than would otherwise be expected. This warming is reduced by the greater cloudiness

Fig. 5.41. Schematic of the
accentuation of the
monsoon by a raised land
surface.

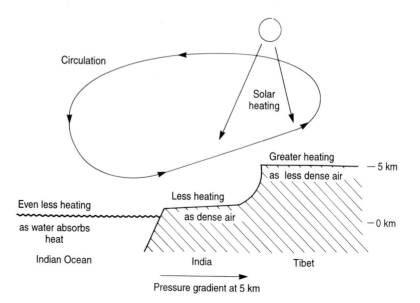

Fig. 5.41. Schematic of the accentuation of the monsoon by a raised land surface.

(and hence more reflection of incoming solar radiation) of Tibet and the Himalayas than would be expected in the absence of these mountains. However, the increased release of latent heat through condensation within the clouds will partially compensate for this effect.

This amplified warming at 4.5 km into the atmosphere over Asia will lead to a greater horizontal pressure gradient between the land mass and the Indian Ocean. This magnification of the pressure gradient is necessary to balance the Coriolis force, induced as the oceanic air moves landward, in such a way as to allow advection into India, rather than the westerlies that would otherwise occur. In §6.1.3 we will see how the uplift of the Tibetan plateau some 5–20 million years ago apparently triggered the onset of the Indian Monsoon.

5.2.7   *The extra-tropics and other tropical air–sea interaction*

The teleconnections of the Southern Oscillation are substantial, and the strongest direct link driven by ocean–atmosphere interaction between the tropical and extra-tropical climate. However, a number of other less dramatic bonds are worthy of mention.

Hurricanes were seen in §2.11.1 to be dramatic, although short-lived, atmospheric features driven by oceanic fluxes. These storms occasionally penetrate into mid-latitudes in weakened form, notable in the northwest Pacific and Atlantic Oceans. More common is the blending of the energy and moisture of a dying hurricane with a pre-existing or developing mid-latitude depression. The resulting cyclone can display particular vigour and the anomalous heating, particularly aloft, due to the warm and moisture laden air can remain for some distance. Most years both the United States and western Europe experience one or two such systems during the summer.

Another mechanism for transporting moist tropical air, ultimately of oceanic origin, to the extra-tropics is the presence of pronounced vertical shear in the sub-tropical atmosphere. The Hadley cells will push warm, moist tropical air polewards at height (Fig. 1.5). In particular synoptic situations coherent pockets of such air can extend far from the generating convergence zones. If this air is uplifted by sliding over drier air at lower altitudes then considerable rainfall can result. Australia occasionally experiences such systems, leading to flooding of the normally dry salt lakes in the continental interior such as occurred in January 1974.

Further reading

A complete reference list is available at the end of the book but the following is a selection of the best books or articles to follow up particular topics within this chapter. Full details of each reference are to be found in the Bibliography.

Bigg (1990): A simple introduction to El Niño.
Hartmann (1994): Excellent discussion of basic meteorological elements of climate and physical feedbacks.
Lamb (1977): Now a somewhat old text but a very comprehensive coverage of climatic history with particular attention to the interaction of the ocean and climate in the North Atlantic sector.
McIlveen (1992): Good discussion of basic meteorology.
Philander (1990): A full discussion of tropical oceanography and meteorology, focussing on El Niño. More mathematical but the first couple of chapters give a very readable introduction.
Tomzcak and Godfrey (1994): A very thorough survey of both the descriptive and dynamical aspects of oceanography, concentrating on regional variations.
Wallace and Gutzler (1981): One of the foundation papers on relationships between the NW Atlantic and climate over western Europe.

6 *The ocean and natural climatic variability*

The history of our planet is one of climatic change. Geology records the effect of the environment – including climate – on tectonic and geomorphological processes. More recently, the evolution and dispersion of humanity, the development of centres of early civilization, and the rise and fall of cultures have been significantly affected by climatic change.

The oceans have been a crucial, if neglected, part of this history. In the previous chapters we have seen how the ocean and climate are linked in many ways. This chapter will explore how these mechanisms have influenced climatic variability on the varied timescales of Earth's history. For eons plate tectonics determined the basic climate (§6.1). Over the past million years or so cycling in the distribution of solar radiation over the Earth's surface has led to dramatic oscillation in the characteristics of terrestrial and marine environments (§6.2). Since the end of the last glaciation significant variability in climate has occurred, much of it affected by air–sea interactions (§6.3), and even in the last 100 years such variability is noticeable (§6.4). This recent change may be driven by anthropogenic alteration to the environment; Chapter 7 will examine the likely impacts of this new forcing term in our globe's climatic equation and consider whether such predicted changes are yet detectable.

6.1 The oceanic role in the geological evolution of climate

Strong maritime influence on global climate has been the norm over the past 570 million years. Earlier, in the Pre-Cambrian (see Fig. 6.1 for a geological time chart), climatic evidence is much more sparse. Theories of stellar evolution suggest that the Sun should have been substantially fainter in the first few billion years of Earth history. Only sporadic evidence for ice sheets suggests, however, that the planet was <u>not</u> significantly colder. Current belief is that a strong greenhouse effect, driven by CO_2 and H_2O, compensated for the reduced solar input to the climate.

6.1.1 *The Palaeozoic and early Mesozoic*

Geological evidence over present-day continental interiors indicates that marine conditions were widespread for much of the period 570–100 million years Before Present (BP). Large parts of North America and Eurasia were under shallow seas with mainly non-glacial climates dominating, apart

176

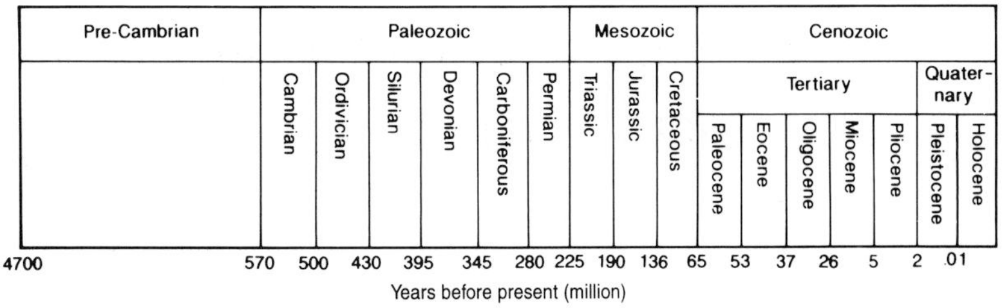

Fig. 6.1. Geological time chart.

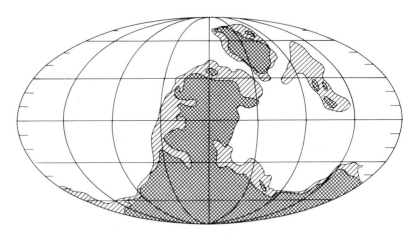

Pre-Cambrian	Paleozoic						Mesozoic			Cenozoic							
										Tertiary						Quater-nary	
	Cambrian	Ordovician	Silurian	Devonian	Carboniferous	Permian	Triassic	Jurassic	Cretaceous	Paleocene	Eocene	Oligocene	Miocene	Pliocene		Pleistocene	Holocene
4700	570	500	430	395	345	280	225 190		136	65	53	37	26	5		2	.01

Years before present (million)

Fig. 6.2. Reconstruction of global geography during the Late Carboniferous (300 million years BP). The cross-hatched regions are dry land and the singly-shaded areas show continental shelves. Data from Parrish et al. (1986).

from two major glacial epochs 470 million and 300 million years BP. Unlike more recent changes in sea level (§§6.3, 6.4) the high marine stand was probably due to variability in tectonic activity.

The geographical distribution of land and sea was seen in Chapter 5 to be of crucial importance to both regional and global climate. Flooding and drying of extensive land areas will lead to climatic perturbation (§6.1.2) but more important is the continental configuration, determined by purely tectonic forces. Just as today's global climate is dominated by the two super-continents of Eurasia–Africa and the Americas, past climates were influenced by other combinations. For much of the Palaeozoic global geography was dominated by the the movement of the super-continent Gondwanaland, and its break-up. The palaeogeography of 300 million years BP is shown in Fig. 6.2. This was during the Carboniferous glaciation, brought about by the large land-mass centred over the South Pole at that time (contrast this with Fig. 1.25, the palaeogeography of the mild Cretaceous, at 100 million years BP). Maritime influence is seen during the Carboniferous, however, in the mild, damp, coal-forming climate of North America and Europe. These continents, while in relatively high latitudes, were then in a Northern Hemisphere devoid of polar land masses. A polar continent cools strongly in winter, sufficient for ice caps to form, leading to a cooling feedback from reflection of incoming radiation. In contrast, while sea-ice may form in a polar sea during winter, an ocean basin with both polar and tropical seas is likely to have sufficient poleward heat transport, and adequate local heat capacity, to melt such ice during the summer, even if one pole is glaciated.

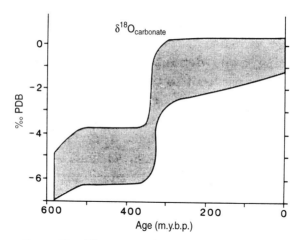

Fig. 6.3. Temporal evolution of the $\delta^{18}O$ ratio in sediments derived from marine carbonates. The two curves show the range of data recorded for different time slices. The distinct discontinuity around 350 million years BP exceeds the usual data range. [After Veizer *et al.* (1986). Adapted from *Geochem. Cosmochim. Acta*, **50**, 1679–96, copyright (1986) with kind permission from Elsevier Science Ltd., The Boulevard, Langford Lane, Kidlington OX5 1GB.]

Some 50 million years previous to this glaciation a dramatic change in oceanic conditions is suggested by the geochemical record. Evidence for increased oxygenation of deep waters (through lower organic carbon content in marine deposits) implies the initiation of deep convection, presumably as an increasingly polar Gondwanaland led to strong cooling of the surface waters of coastal seas by cold southern winters. At the same time the proportion of oxygen isotopes in marine fossil skeletons changed dramatically. The ratio of the relatively rare ^{18}O isotope to common ^{16}O in the carbonate-rich shells of particular species of marine plankton is related to the environment in which the plankton live. The organisms use sea water and dissolved carbonate in the formation of their hard skeletons; the chemical and physical processes by which oxygen atoms from this water enters into the shell material are moderated by the water's temperature. This isotope ratio is expressed as a difference, $\delta^{18}O$, from the ratio in a standard[1] sea water sample:

$$\delta^{18}O = \left(\frac{^{18}O/^{16}O_{sample} - {}^{18}O/^{16}O_{standard}}{^{18}O/^{16}O_{standard}} \right) \cdot 1000 \qquad (6.1)$$

A sudden increase in $\delta^{18}O$ of about 4‰ accompanied the onset of deep convection (Fig. 6.3). From modern experiments it is known that an increase in environmental temperature of 1°C lowers the $\delta^{18}O$ of plank-tonic shells by about 0.2‰ (that is, ^{16}O atoms are preferentially incorporated into the material, at the expense of ^{18}O). In the absence of change in other environmental factors, Fig. 6.3 suggests a sudden fall in temperature of 20°C. This almost certainly overstates the real global temperature change. Apart from the uncertainty of knowing whether palaeo-plankton obeyed the same temperature–isotope relationship as present-day species, and the unlikely chance that these few data reflect global rather than regional change, alteration of the background $\delta^{18}O$ is also possible. Changes in salinity alter $\delta^{18}O$: ^{16}O evaporates at a faster rate than ^{18}O,

[1] The concept of a 'standard' is common throughout chemistry. It denotes a given, readily reproducible mixture against which samples taken from the environment can be contrasted. In this case the sample is sea water of a given oxygen isotope ratio, taken from the tropical ocean.

Fig. 6.4. Schematic illustration of processes which can modify the ^{18}O content of sea water.

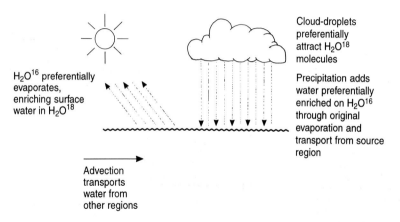

H_2O^{16} preferentially evaporates, enriching surface water in H_2O^{18}

Advection transports water from other regions

Cloud-droplets preferentially attract H_2O^{18} molecules

Precipitation adds water preferentially enriched on H_2O^{16} through original evaporation and transport from source region

leading to enrichment of the remaining water in ^{18}O. The formation of ice sheets also enriches oceanic water in ^{18}O, because of the removal from the hydrological cycle of large amounts of precipitated water, rich in ^{16}O because of the evaporation mechanism and the preferential incorporation of ^{18}O in cloud droplets from the vapour phase. Advection of water from regions with different water balances, and hence background $\delta^{18}O$, can also alter the local sea water oxygen isotope balance. These processes are summarised in Fig. 6.4. The true significance of this sudden change in $\delta^{18}O$ is therefore not known; an added complication is the synchronous evolution of terrestrial plants leading to a more oxygen-rich atmosphere, perhaps with a new isotopic composition.

The southern continent of Gondwanaland (see Fig. 6.2) was just the first stage of the formation of the truly super-continent of Pangaea (see Fig. 1.25). By about 250 million years BP all the Earth's land seems to have formed one large mass stretching from the South Pole to the North, although without extensive polar land masses in either hemisphere. Both geological and modelling evidence suggest that strong western boundary currents washed the eastern shores of Pangaea, transporting warm water poleward to keep polar ice sheets at bay and also to warm the east coast relative to the west. Continental interiors, nonetheless, would have remained subject to strong seasonal extremes.

6.1.2   *The Cretaceous: a case study*

The evolution of the Earth's climate over the last 200 million years has largely been a consequence of the break-up of Pangaea. This gave rise to the formation of the Atlantic Ocean, the temporary creation of the globe-spanning tropical Tethys Sea, movement of the Antarctic plate towards the South Pole and the plates of modern Africa, India and Australia northwards. It is also a period about which we have increasing information. Much effort has been invested in studying the mid-Cretaceous, about 100 million years BP, as it possessed the most recent warm, (probably) ice-free global climate, and is prior to the suspected asteroid impact around 65 million years BP and the mass extinction of species leading to the eventual dominance of mammals. Here we will examine the oceanic influence on climate in this epoch.

The land distribution of this period is shown in Fig. 1.25. There were significant differences from modern geography. Sea level was higher, flooding an additional 15–20% of the land surface. This was due to a combination of tectonic and physical (that is, expansion and ice melt) contributions. The blocking of seaways in the mid-latitudes of both hemispheres resulted in pronounced western boundary currents transporting heat polewards in the ancient Pacific. The Tethys Seaway, connecting all tropical oceans, allowed marine communication between different oceans, although modelling experiments suggest a gyre circulation within the Seaway, rather than an equatorial-like westward flow. The limited amount of polar land will have also contributed to warmer climates. However, neither oceanic transport nor land distribution is sufficient to explain the apparently mild polar regions of the geological record. Currently it is believed that the atmospheric CO_2 level must have been anything from 2 to 10 times greater in the Cretaceous compared to today. Definite evidence for this is, however, lacking.

The oxygen isotope record suggests intermediate-deep waters were some 15°C warmer than today, with deep water being warm and saline, originating from a strongly evaporitic environment in the sub-tropics. While such conditions are inconsistent with polar winter cooling and deep convection the geological evidence is still unclear. Known changes to $\delta^{18}O$ from non-temperature effects (§6.1.1) also suggest caution is required in interpretation of such data.

Sea surface temperatures also suggest caution is warranted. While mid- and high latitude sea surface temperature estimates are higher than for present day, tropical area values are not significantly so. The greater supply of latent heat to the atmosphere, both from a greater geographical area of ocean from which evaporation can occur, and the higher saturation vapour pressure resulting from higher mid- and high latitude temperatures, led to greater storminess in some model simulations. This would assist deep convection.

6.1.3   Tertiary climates

Tertiary climate is characterised by a slow trend from global mildness to bipolar ice sheets, with abrupt transitions accompanying the onset of new steps in this evolution. A decrease in atmospheric CO_2 is thought to have occurred during this period, but much of the climatic change will have occurred due to tectonic variation in the Earth's geography. This has been dominated by the continued break-up of the Cretaceous remnants of Pangaea, and the formation of the new supercontinents of Eurasia–Africa and the Americas (Fig. 6.5). In the process the Tethys Seaway was closed (30 million years BP), the Alps (20 million years BP) and Himalayas (15 million years BP) were uplifted by the collision of Africa and India respectively with Eurasia, the Southern Ocean was opened by the northward movement of

Fig. 6.5. Palaeogeographic maps at various times during the Tertiary; (a) the middle Palaeocene (65 million years BP), (b) the middle Oligocene (33 million years BP), (c) the middle Miocene (15 million years BP), (d) the early Pliocene (4 million years BP). Light coloured areas show flooded continental shelves, oceanic ridges are shown in black, hypothesised surface flow is shown by thin arrows with deep flow having thick arrows. [From Haq and Van Eysinga (1987). Reproduced with kind permission of Elsevier.]

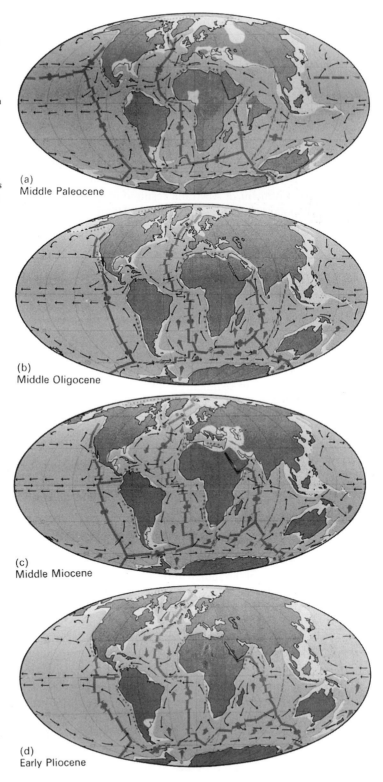

(a) Middle Paleocene

(b) Middle Oligocene

(c) Middle Miocene

(d) Early Pliocene

Australia (20 million years BP), and the Panama Isthmus was sealed (4 million years BP).

The trend in global climate through the 60 million years of the Tertiary was not linear, or even monotonic (see Fig. 1.24). During the Eocene (50–55 million years BP) global air and sea temperatures increased to their highest values since the Cretaceous. This may have occurred due to increased tectonic vigour, releasing CO_2 into the atmosphere through volcanic activity, or hydrothermal processes along the mid-Atlantic Ridge adding calcium to the deep ocean. The latter allows enhanced marine precipitation of calcium carbonate, forcing CO_2 into the ocean, and eventually the atmosphere (see §3.3.2, equation 3.10). It is also likely that the northward movement of Australia resulted in greater transport of tropical waters to polar latitudes of the Southern Hemisphere, and that the growing width of the Atlantic encouraged the development of western boundary currents. Both of these ocean circulations would result in enhanced poleward heat transport.

A significant cooling occurred over the next 10 million years (Fig. 1.24), which may have been associated with the onset of Antarctic glaciation. The reasons for this cooling are unknown; certainly the palaeogeography was still compatible with enhanced poleward heat transport (Fig. 6.5). Another, abrupt, cooling occurred around 34 million years BP. This is associated with finds of ice-rafted debris in Antarctica, suggesting the presence of sea-ice and perhaps coastal glaciers. Changes in $\delta^{18}O$ of plankton from concurrent sediments in the equatorial Pacific support a rapid increase in ice volume. In contrast to the earlier cooling, the most noticeable contemporary temperature change appears to be in the Northern Hemisphere. It is thought that limited deep water formation in the North Atlantic may have begun at this time. A further significant increase in Antarctic ice volume probably occurred about 30 million years BP.

The gradual alignment of the land plates into their modern positions resulted in another abrupt global climate shift around 10–15 million years BP. Australia had by this time moved sufficiently far north that a definite Circumpolar Current was created, isolating Antarctica climatically from the warmer conditions further north. A major increase in glaciation probably resulted. This period also saw the major Tibetan uplift from the impact of the Indian and Eurasian Plates. This had a significant impact on the atmospheric circulation, but for our discussion its principal effect was the initiation of the Indian Monsoonal circulation. Deep water formation in the Atlantic probably approached modern levels somewhat later, about 7 million years BP, while a temporary closure of the Indonesian Seaway about 8 million years BP would have created enhanced poleward heat transport in the North Pacific.

The prelude to widespread, oscillatory glaciation in the Northern Hemisphere was climatically chaotic. Most of this variability must have been due to internal dynamics in the climate system as mostly minor tectonic adjustments remained before current geography was attained. Around 5.5 million years BP a rapid extension of ice cover over Antarctica lowered sea level some 50 m. This seems to have isolated the Sea of Japan,

Fig. 6.6. Difference in poleward ocean heat transport (in 10^{15}W) between an ocean model simulation with the Isthmus of Panama flooded and one with present-day geography. Note that the poleward heat flux is significantly reduced globally when there is a tropical connection between the Pacific and Atlantic, with the major effect in the Atlantic. [Reprinted, with kind permission, from Fig. 10c of Maier-Reimer and Mikolajewicz, *Paleoceanogr.*, **5**, 349–66 (1990), copyright of the American Geophysical Union.]

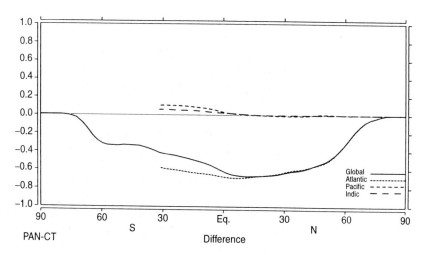

and in combination with uplift associated with the tectonic clash of Africa with Europe the Strait of Gibraltar was drained, also isolating the Mediterranean. Excess evaporation over this basin would have dried it completely in a few hundred to thousand years. The depth of the salt deposits on the Mediterranean floor suggest that this drying, followed by a flooding, must have occurred perhaps 40 times. Such a loss of dissolved salts would have led to a decrease in the average salinity of the global ocean, perhaps by 2‰. This would have significantly altered the basic density structure, especially in the polar regions. It is known that the thermohaline circulation is sensitive to quite small salinity perturbations, so the processes and locations of deep water formation are likely to have been significantly different prior to the Mediterranean drying cycles.

Some evidence of Alaskan glaciation around 5.5 million years BP has been found but substantial warming, accompanying a halving of the Antarctic Ice Sheet, set in between 4 and 5 million years BP. This may have been partially due to the closure of the Panama Isthmus, which would have enhanced oceanic heat transport poleward in the Atlantic (Fig. 6.6). This warm period did not last. By 3.5 million years BP renewed glaciation in Antarctica and Patagonia is hypothesised. Over the next million years evidence for Northern Hemisphere glaciation in Iceland, Greenland and the Rocky Mountains becomes increasingly evident. This cooling trend was particularly sharp around 2.5 million years BP. Convection in the Norwegian Sea is also evident from this period.

The sequence of abrupt coolings and warmings during the late Tertiary are not dominated by astronomically forced periodicity, as in the Quaternary (§6.2). The abruptness of climatic change must therefore be linked to instabilities in the climate system. Rapid changes in oceanic and atmospheric CO_2 levels are possible due to change in the rate of tectonic movement. Ice sheet feedbacks (§1.4) can also lead to rapid accentuation of change. Similarly, very modest changes to the oceanic water balance can have profound impact on the thermohaline circulation, even to the extent of changing its essential character. We will encounter some such changes in

Fig. 6.7. Composite record
of $\delta^{18}O$ fluctuations for the
last 2.5 million years, based
on three deep-sea cores.
[Reprinted from Crowley
and North,
Paleoclimatology (1991)
with kind permission of
Oxford University Press.
Based on data from Raymo
et al. (1990).]

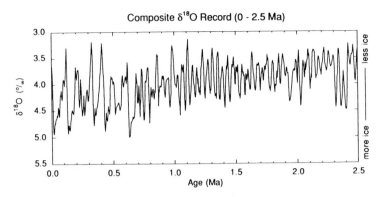

Fig. 6.7. Composite record of $\delta^{18}O$ fluctuations for the last 2.5 million years, based on three deep-sea cores. [Reprinted from Crowley and North, *Paleoclimatology* (1991) with kind permission of Oxford University Press. Based on data from Raymo *et al.* (1990).]

later sections of this chapter, brought about by much smaller perturbations to the hydrological cycle than occurred in the Tertiary.

6.2   The ocean and Quaternary glaciation

The magnitude and geographical distribution of solar radiation is the basic source of energy for the climate system. In §1.8 we saw that the Earth's orbit fluctuates over time, with three well defined periodicities arising from different perturbations to the orbit (Fig. 1.27). The energy variation associated with these perturbations has dominated the global climate during the last two million years (the Quaternary). Variation in ice volume, and therefore global temperature, over this period is shown in Fig. 6.7, using $\delta^{18}O$ from three deep sea cores as a proxy indicator (see §6.1.1). A general increase in $\delta^{18}O$ over time – due to an overall rise in ice extent – is overlain by a high amplitude, geologically fast family of oscillations. Prior to about 700 000 years BP the periodicity of this oscillation is roughly 40 000 years, corresponding to pulsation in the Earth's obliquity, or tilt. More recently, the principal frequency has flipped to the longer, 100 000 year cycle in eccentricity, or the degree of circularity of the Earth's orbit. The variation in the three orbital parameters over the last 800 000 years is shown in Fig. 6.8. The two shorter period fluctuations have a regular amplitude swing, unlike the eccentricity (although note the strong beating in precession). The basic shape of the Earth's orbit is strongly modulated by period variation even longer than 100 000 years. This contrast in forcing amplitude is mirrored in the variability of $\delta^{18}O$ before and after the 700 000 years BP discontinuity (Fig. 6.7).

While a clear link between astronomical forcing and climatic fluctuation exists three major puzzles arise. Why did this forcing become dominant only 2 million years ago? What caused the break in character of the climatic record at 700 000 years BP? Why has the more recent eccentricity phase of climatic variability been of greater amplitude than the previous obliquity phase, as the latter provides a (substantially) bigger fluctuation in solar forcing? The solution to each of these lies within the climate system, and marine influences may well be significant.

In the previous section we saw how predominantly tectonically-driven changes led to the formation of ice sheets during the late Tertiary. The

Fig. 6.8. Variation in eccentricity, obliquity (in degrees) and precession over the past 800 000 years (based on Berger, 1978), with the normalised combination of these giving the total curve, labelled ETP. The variance spectra are shown to the right, with the dominant periods, in thousands of years, indicated. [Fig. 2 of Imbrie *et al.* (1984) in *Milankovitch and climate.* Reproduced with kind permission of Kluwer Academic Publishers.]

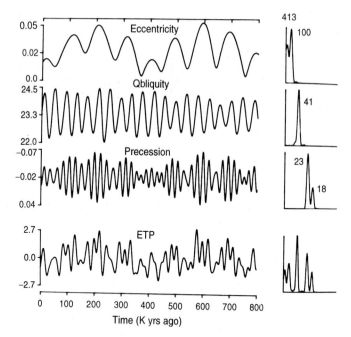

geologically-speaking high frequencies of astronomical forcing are observed in the climatic record of these times (and indeed even earlier). However, the temporal resolution of the geological record, and the absence of extensive Northern Hemisphere glaciation, make this signal both difficult to detect and probably of smaller amplitude. The presence of ice lends a distinctly unstable element to climate. Positive feedbacks relying on the loss of energy from reflection of much incident radiation produce cooling and further ice growth. Other positive feedbacks arising from ice melting and so more radiation absorption give warming and further ice retreat. These feedbacks are particularly pronounced over the extensive land masses of the Northern Hemisphere; in Antarctica the Southern Ocean effectively limits the extent of glaciation. Sea ice is much less easy to maintain because of the constant heat flux from the unfrozen water beneath. Therefore rapid changes in global climate are more likely to occur once glaciation has a foothold in Eurasia and North America. This began in the late Tertiary.

The most extensive glaciations all occur as part of the more recent, 100 000 year cyclicity phase of the climate. The reason for this increase in amplitude, but decrease in glacial frequency is still the subject of active research. The geographical variation of solar radiation provided by the obliquity signal seems more important for Northern Hemisphere glacial control, as when the tilt increases less radiation reaches the Northern Hemisphere during northern summer, allowing snow and ice to remain longer at lower latitudes. The eccentricity variation, in contrast, determines through the year the net global radiation received by the Earth: a low eccentricity means a similar basic energy supply at all times, while a high eccentricity leads to perhaps 2% variation through the year.

The change in periodicity must mean that the climate system passed a threshold of some kind, leading to more extensive, and so more persistent, glaciation. Several mechanisms have been suggested. The discontinuity in periodicity occurs when the eccentricity was low (c. 750 000 years BP, Fig. 6.8) and the precession was minimal. A high obliquity at this time may have encouraged ice sheet growth at a time of weak summer warmth due to the small eccentricity. If this growth was sufficient to spread Northern Hemisphere glaciation far enough south to survive the next low obliquity phase (because of albedo feedback), then a net increase in minimum glaciation may have occurred. Unfortunately, the extent of this glaciation, as with all those prior to the last one, is difficult to define well, because much evidence was destroyed by subsequent advances.

Other possible reasons for the change in glacial periodicity are linked to the impact of the Tibetan plateau on the atmospheric circulation. A significant amount of the uplift in this region, perhaps more than 1 km, may have occurred during the early Quaternary. Uplift in other areas, such as the mountains of western North America, also may have been significant during this time. Higher land masses tend to increase the Earth's albedo both because of the greater area of winter snow and more reflective alpine vegetation, but also because there is less atmospheric absorption. Crossing a threshold albedo may have pushed the planet into a more glacially attractive state, breaking the 40 000 year cycle. Another mechanism linked to the Tibetan and North American uplift history derives from the distorting effect of high mountains on atmospheric storm tracks. The Rockies induce southward shift in the storm-steering tropospheric planetary waves (§5.1.4 and Appendix D), both over eastern North America and western Europe. Tibet may amplify this signal. Shifting storm tracks southward increases winter snowfall and so promotes glacial growth.

Oceanic conditions also promote winter snowfall. The strong western boundary currents of the North Atlantic and Pacific, while supplying heat poleward and potentially discouraging glaciation are also responsible for enhanced storm generation (§5.1.1). This, coupled with the southerly penetration of the upper level steering flow discussed above, may enhance the growth of glaciers. The actual trigger for the discontinuity in glacial periodicity is likely to be some combination of these, and other, processes.

6.2.1 *Interglacial termination*

The climatic record of the Quaternary shows oscillations between two quasi-stable states: glacial and interglacial. Modelling experiments lend support to this idea through the existence of two stable states for the oceanic thermohaline circulation. However, detailed inspection of the climatic record over the last glacial cycle (Fig. 6.9) shows a less clear signal. Interglacial periods, such as the last 10 000 years and the warm interval around 125 000 years BP, are distinctly different climatically. In §6.2.2 we will discuss the rapid process of de-glaciation but first we consider the slower evolution of glaciation.

Interglacial periods have basically similar climates to those we are

Fig. 6.9. $\delta^{18}O$ record in remains of *Globigerinoides ruber* from a core in the equatorial Indian Ocean. High values of $\delta^{18}O$ correspond to low ice volume or interglacial periods. [After Fig. 2b of Rosteck *et al.* (1993). Adapted with permission from *Nature*, **364**, pp. 319–21. Copyright 1993 Macmillan Magazines Limited.]

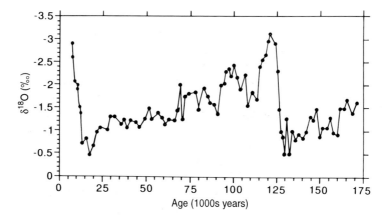

familiar with today. However, significant variations about this state are possible. The last interglacial is believed to have been 1°–3°C warmer than today. Sea level was some 6 m higher, presumably because of smaller Greenland or West Antarctic Ice Sheets. Climatic zones in several parts of the world were displaced 10° poleward, in comparison with today. The Indian Monsoon was 50% stronger, due to the greater heating of the Tibetan Plateau. A stable climate may, nevertheless, have been absent. Recent evidence from ice cores in Greenland suggests that marked variability may have been experienced during this inter-glacial regime.

The shift to glacial conditions is also subject to variability. While a rapid cooling occurred between 120 000 and 110 000 years BP (Fig. 6.9) two periods of warmer climate preceded the real onset of glacial conditions around 75 000 years BP. Even then, a milder phase from 60 000 to 30 000 years BP occurred before the peak of the last glaciation, around 20 000 years BP.

The onset of glaciation is still not well understood. A phase of the Milankovitch cycle with cool summer orbits would permit ice to accumulate more readily in sub-polar latitudes, and so allow existing ice sheets to grow or new ones to form. The former seems to be the likely mechanism for providing sufficient feedback to progress beyond the initial cooling. However, water balance and snow accumulation parameterisations in climate models are not currently reliable enough to allow such theories to be confidently explored numerically.

Accompanying the decline in temperature was a fall in atmospheric CO_2 and methane (Fig. 6.10). These decreased by a third and a half respectively. The links between such alteration to greenhouse gases and climatic change are hotly debated. During part of the last glacial cycle variations in these greenhouse gases and local temperature were synchronous (10 000–40 000 years BP); at other times the trace gas changes appear to lag the climate (110 000–120 000 years BP). The degree of correspondence also depends on where ice cores are sampled: the Antarctic record shows greater synchronicity than Greenland. Most theories for the decline in atmospheric CO_2 assume that the ocean acts as the carbon reservoir. One theory, proposed by Broecker, is that falling sea level allows greater erosion of continental

Fig. 6.10. Variation in CO_2 and CH_4 over the past 160 000 years, from ice-core records. [Adapted from Houghton *et al.* (1990), with kind permission of Cambridge University Press.]

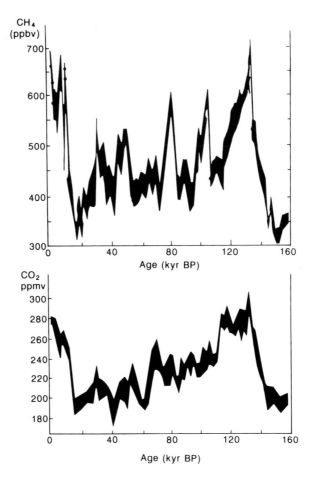

shelves. The nutrients so released into the ocean encourage greater biological productivity and synchronous draw-down of atmospheric CO_2. This mechanism would account for a lagged atmospheric response, but proxy observations of past oceanic nutrient levels point to the need to invoke other mechanisms to explain the size of the decline.

Most other mechanisms proposed to explain decreased glacial CO_2 also rely on a lagging of climatic changes. The glacial ocean is characterised by much weaker deep water formation. This occurs because of a relative decrease in evaporation, and hence surface salinity, in the sub-tropics as the climate cools. The water supplied to the northern Atlantic by the Gulf Stream is therefore less dense and overturns less readily. Modelling experiments have shown that the thermohaline circulation in the Atlantic is quite sensitive to such salinity forcing. Copious deep water formation, as occurs at present, could be drastically cut by changes of surface salinity well within expected limits.[2] Once deep water formation was lessened nutrients in high latitudes may be better utilized. Productivity, plus CO_2 draw-down,

[2] It should be noted, however, that an increase in ice volume raises salinity throughout the ocean, thus making sea water at all depths denser than today. At the last glacial maximum sea level was 120 m lower than today; this implies an increase in oceanic salinity of about 1 ‰.

would be enhanced. However, deficiencies in light, iron and present day nutrients (§4.1.1) lessen the probability of this mechanism.

Another way of decreasing glacial CO_2 is by enhancing productivity through strong upwelling of nutrients. In some areas of the glacial ocean upwelling may have been enhanced, such as in the tropical Atlantic, because of inferred stronger Trade winds. High productivity is associated with siliceous diatoms as opposed to carbon-based plankton (§4.1.1). Such a change to the dominant plankton species would raise the upper ocean carbonate concentration, and, from equation (3.3), draw more atmospheric CO_2 into solution. However, a reduction in upwelling in the glacial Indian Ocean by up to 90%, because of a weaker Indian Monsoon, would probably counter much of this effect.

The decrease in methane is also a complex and controversial process. The natural source for methane is predominantly anaerobic decay in wetlands (§4.2.1); the ocean is likely to contribute at most only a few percent to the signal. Decreases in high latitude wetland area associated with increased ice volume would cause a lagged decline in CH_4, as would dry periods in tropical climates.

Glaciation, in summary, is likely to be driven by astronomical forcing but enhanced by ice, oceanic and atmospheric feedbacks. The latter are highly non-linear and cause the irregular nature of the change from interglacial conditions.

6.2.2 *Glacial termination*

Deglaciation is a very rapid process, in geological terms. Both the present and last interglacials were established only a few thousand years after peaks in the previous glacial period (Fig. 6.9). There is some evidence that large steps in deglaciation can occur over perhaps only a few decades. Despite the strong, stabilising ice albedo feedback, once an ice sheet begins to melt overwhelming positive feedbacks must reinforce its retreat.

The trigger for deglaciation is presumably astronomical, but this is not unequivocable. Increased glaciation leads to a decline in the global precipitation: once insufficient snow accumulates each winter to replace that melted in the following summer deglaciation may occur of its own volition. It is suggestive that maximum global ice volume immediately precedes every glacial–interglacial transition in the Quaternary.

Despite its rapidity, deglaciation is not monotonic. During the last deglaciation a notable re-cooling – the Younger Dryas – occurred about 12 000 years BP (Fig. 1.28). This appears to have been a global halt in warming, although its strongest signal is around the shores of the North Atlantic. This ocean, and its interaction with the atmosphere, almost certainly provides the key to the speed of deglaciation. However, we need first to set the background by briefly describing global climate at the peak of the Last Glacial Maximum (LGM), about 20 000 years BP.

The extent of ice in the Northern Hemisphere at the LGM is shown in Fig. 6.11, with estimated surface temperatures during July in Fig. 6.12. Much of North America north of 48°N and Europe north of 52°N was

Fig. 6.11. Reconstructed
ice-cover of Northern
Hemisphere at 21 000 years
BP. Data from Peltier
(1994).

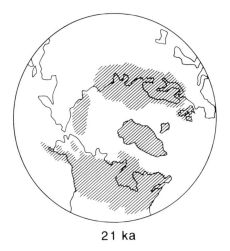

21 ka

Fig. 6.12. Surface air
temperature in July at the
LGM, in °C. From
CLIMAP and atmospheric
model data (courtesy of
Paul Valdes).

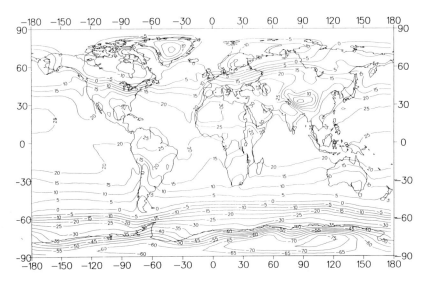

covered in ice up to several kilometres in thickness. Extensive mountain
glaciers existed in the Alps and Pyrenees, with winter temperatures over
Europe, the North Atlantic and North America probably several tens of
degrees colder than today. Sea-ice may have extended in winter as far south
as northern Spain, although recent palaeoceanographic and modelling
evidence is beginning to cast doubt on such extensive coverage. An
alternative view is that even in the Norwegian and Labrador Seas there was
no permanent sea-ice during much of the last glaciation. Elsewhere, Alaska
and Siberia were only partially ice covered. Similarly, in the Southern
Hemisphere there was glaciation in New Zealand and Patagonia, but little
extension of the Antarctic Ice Sheet compared to today. The globe was, on
average, drier at the LGM than today.

The ocean circulation was markedly different to today. The Gulf Stream
system in the Atlantic re-circulated near 40°N, not penetrating into the NE
Atlantic as at present. Deep water formation in the Atlantic was probably

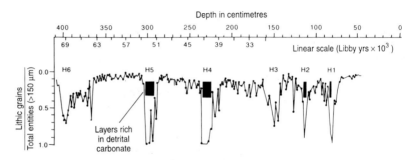

Fig. 6.13. Proportion of lithic grains (sediments of land origin) in large grained sedimentary material at a core in the North Atlantic. Large increases in this proportion occur when iceberg calving increases; 6 such events over the past 70000 years are shown, labelled H1-H6. [After Fig. 2a of Bond *et al.* (1992). Adapted with permission from *Nature*, **360**, pp. 245–9. Copyright 1992 Macmillan Magazines Limited.]

drastically curtailed, and either shifted south off the west European coast or into the Labrador Sea. Some deeper water was now supplied by enhanced subduction in the sub-tropical gyres. The global thermohaline circulation was therefore much slower; deep water took perhaps 3000–6000 years to return to the surface, rather than 1000 or less, as today. The flow of deep water remained, however, into the Pacific.

This basin in general experienced relatively little change. The Kuroshio probably moved south during winter, but seems to have been displaced only slightly in summer (Fig. 6.12). There is evidence, however, that increases in evaporation in the Pacific may have been sufficient to allow some deep water formation in the North Pacific. Such salinity increases would have been preserved by the closure of the Bering Strait through lowered sea level. Other oceans also showed distinct contrasts to today. The Indian Monsoon was much weaker, leading to a reduced seasonal oscillation in surface circulation in the Indian Ocean, and summer upwelling of only 10% of present values. The western basin of the Mediterranean was much colder than today, while retaining its high salinity. Model results suggest that while the volume of water exchanged with the Atlantic was roughly similar to today (perhaps halved), its origin was then in the western basin. In the simulation the eastern basin did not contribute to the exchange because the cold, dense water of the west provided the deep water for the eastern basin, in contrast to today. The Mediterranean outflow would therefore have been 7°–8°C colder than today and not have contributed so significantly to intermediate level circulation in the Atlantic.

A striking feature of the North Atlantic during glacial periods is the periodic discharge of enormous numbers of icebergs from the eastern North American, or Laurentide, ice sheet (Fig. 6.13). These are known as 'Heinrich' events. They occur at the end of periods of cooling, roughly every 10000–15000 years, and are followed by abrupt shifts to warmer conditions. It is not known whether these events are linked to orbital forcing or internal ice sheet dynamics. However, the latter is probably more likely. If the extent and gradient of the Laurentide ice sheet exceeded some dynamic stability constraint a slip, analogous to those resulting in avalanches or mud-slides but on a much greater scale, could be imagined that may lead to catastrophic breakage. Once a significant fraction of the ice sheet was lost the weakening of the albedo feedback would induce an immediate warming.

Climate remained essentially constant for about 5000 years after the

Fig. 6.14. Reconstructed
ice-cover at 14 000 years
BP. Data from Peltier
(1994).

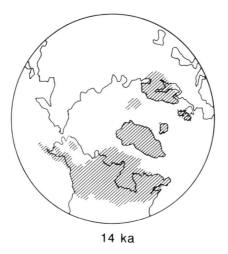

14 ka

Fig. 6.15. Variation of sea
level (*rs*), in metres, over
the last 20 000 years, using
data from Fairbanks (1989)
and Peltier (1994).

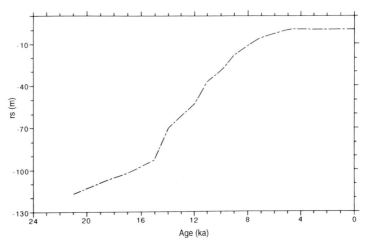

LGM, with an intensification of aridity in Africa, although there is evidence
for ice melt in Antarctica and northern Russia (Fig. 6.14). Sea level rose only
some 10–15 m – a tenth of its LGM depression – by 14 000 years BP.
However, over the next 2000 years sea level rose 60 m (Fig. 6.15) as much of
western North America was deglaciated. This sudden warming coincides
with the last Heinrich event, H_1 (Fig. 6.13). An instability in the shrinking
Laurentide Ice Sheet discharged a large enough proportion of the ice mass
for the ice albedo feedback to be irretreviably weakened. This Heinrich
event is unlike those earlier in occurring during a generally warming phase.
This suggests that the instability may have been due to basal melting. It has
been speculated that sea level rise and associated melting of permafrost at
the beginning of deglaciation may have explosively released methane from
the polar soil. This has been linked to warming through an enhanced
greenhouse effect, although such a notion is difficult to quantify. However,
perhaps such an explosive release could provide a triggering mechanism for
H_1.

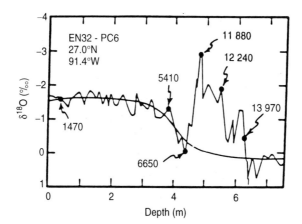

Fig. 6.16. Variation of $\delta^{18}O$ in sediments from a core in the Gulf of Mexico. Radio-carbon dates, in years BP, at selected depths are shown. The very negative excursion of $\delta^{18}O$ from 14 000 to 12 000 years BP comes from the input of very ^{18}O poor melt water, from the North American glaciers. Note that the $\delta^{18}O$ scale is reversed to accentuate this feature. [Reprinted, with kind permission, from Fig. 5 of Broecker *et al.*, *Paleoceanogr.*, **3**, 1–19 (1988). Copyright of the American Geophysical Union.]

Whatever the cause of H_1 a sudden resumption of North Atlantic deep water production followed. The climatic warming allowed the North Atlantic drift to penetrate into the NE Atlantic again, enhancing the warming by providing a heating source to sub-polar latitudes. This led to an acceleration in the melting of the northern European ice sheet, and so further warming. The climate of the NE Atlantic was still cold so the density of the revived supply of warm salty sub-tropical water was readily increased by winter cooling. A thermohaline circulation similar to today's could thereby begin.

The melt water from the Laurentide ice sheet fed into an extensive inland lake system; the present-day Great Lakes network is its remnant. Catastrophic failure of ice walls damming these lakes periodically caused massive flooding, and sometimes sudden injections of fresh water to the ocean. Low $\delta^{18}O$ in Gulf of Mexico sediments shows that much of the melt water initially flowed south via the Mississippi to enter the ocean (Fig. 6.16). Sometime around 11 000 BP melt water was also able to enter the NW Atlantic through the St. Lawrence. Interactions within the climate system then abruptly halted, and partially reversed, the warming (Fig. 1.28). The injections of fresh water into the Gulf Stream led to a reduction in density of the waters of the NE Atlantic, suppressing deep water formation. A pulse of fresh water from the Baltic Lake, in which melt water from the Scandinavian ice sheet had accumulated, is also possible at this time. A further source of water for the North Atlantic around 12 000 years BP is the renewed flow of the fresher North Pacific water through the Arctic as the Bering Strait was flooded (today the latter is only 50 m deep, cf. Fig. 6.15). Turning off the Atlantic thermohaline circulation would decrease poleward heat flow in the ocean, which would, via a reduction in latent heat transfer and associated cyclogenesis, also reduce atmospheric poleward transport. The still extensive ice sheets were then able to stabilise for about 2000 years. This process appears to have consisted of several pulses, with significant changes occurring as rapidly as over just 40 years (Fig. 6.17).

This cooling phase – the Younger Dryas – is well established. However, not all interpretations of deep-sea sediment analyses support the above theory. Deep water formation is monitored through the nutrient level in

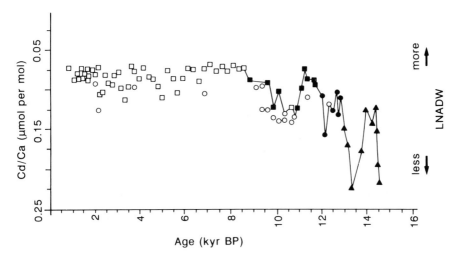

Age (kyr BP)

Fig. 6.17. The Cd/Ca ratio
in benthic foraminifera
from the Bermuda Rise in
the North Atlantic over the
past 15 000 years. Higher
values of this ratio indicate
more nutrients in the deep
water, and can be
correlated with more
quiescent deep flow. [After
Fig. 4a of Lehman and
Keigwin (1992). Adapted
with permission from
Nature, **356**, pp. 757–62.
Copyright 1992 Macmillan
Magazines Limited.]

bottom water. This can be done by examining the Cadmium:Calcium ratio (Cd/Ca) in bottom-dwelling foraminifera that accumulate in the sediment upon death. Cadmium concentration in organisms today correlates well with their phosphorus content. The biochemical process by which cadmium is absorbed into cells in tandem with phosphorus is not known but the relationship is robust. The older a deep water mass, the more organic detritus – and therefore phosphorus-containing compounds – is available for incorporation in foraminifera. The peaks in Cd/Ca shown in Fig. 6.17 therefore should correlate with periods of reduced replenishment of NE Atlantic deep water with low nutrient surface water. Another indicator of productivity is the fractionation of the two isotopes of carbon, ^{12}C and ^{13}C, in plankton. The proportion of the minority isotope ^{13}C absorbed, denoted by $\delta^{13}C$ and evaluated as for $\delta^{18}O$ (equation 6.1), decreases with productivity and therefore increases with the level of oxygenation: more productivity means less of the more readily assimilated ^{13}C is present in solution. Analyses of $\delta^{13}C$ in benthic organisms both in the Norwegian Sea and the NE Atlantic (Fig. 6.18) reveal a less distinct pattern than the Cd/Ca ratio. High $\delta^{13}C$ in the LGM Norwegian Sea suggest local ventilation while low values further south support the lack of outflow from this site. During the Younger Dryas (11–10 000 years BP) the $\delta^{13}C$ record could be interpreted to infer moderate deep water formation and spreading. However the rapidity of downstream response to variations in the degree of deep water formation in the Norwegian Sea is problematic; significant perturbations to the Norwegian Sea record later in the Holocene are not mirrored further south (Fig. 6.18). Another theory, compatible with the above interpretation, ascribes the Younger Dryas cooling to a loss of atmospheric CO_2 from the earlier renewal of deep water formation and associated sequestration of CO_2 into the deep ocean. On balance, however, fresh water anomalies in the surface waters of the North Atlantic cannot yet be dismissed as a trigger for the Younger Dryas. Modelling studies strongly support this mechanism, suggesting a decline in polewards heat transport of 60-70% during strong melt water events.

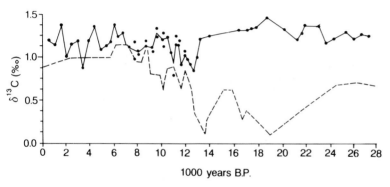

Fig. 6.18. Variation of $\delta^{13}C$ in benthic foraminifera at two sites over the last 25 000 years: solid line – Norwegian Sea (64° 31′N, 0° 44′E), dashed line – NE Atlantic (54° 15′N, 16° 50′W). High values are consistent with recent exposure of the ambient deep water at the surface. [After Fig. 1b of Veum *et al.* (1992). Adapted with permission from *Nature*, **356**, pp. 783–5. Copyright 1992 Macmillan Magazines Limited.]

6.3 The ocean and Holocene climate

The renewed cooling of the Younger Dryas ceased, possibly abruptly, about 10 500 years BP. The origin of the renewed warming is not clear. Several processes leading to such climatic forcing occurred and may have mutually reinforced: an astronomically driven tendency for warmer summers, peaking about 8000 years BP (Fig. 6.8); a resumption of the interglacial thermohaline circulation (Fig. 6.17); and a further increase in atmospheric greenhouse gases (Fig. 6.10). The Laurentide ice sheet persisted for another 3000 years, although essentially modern climatic conditions were established after the last major melt water injection around 9000 years BP (Figs. 6.15 and 6.16).

The relative stability of global climate over the last 10 000 years, reflected in the surface air temperature reconstruction of Fig. 1.28, conceals variation with marked regional amplitude. Much of the globe experienced warmer climate than today – the Climatic Optimum – from 9000 to 5000 years BP (§6.3.1). This was followed by a decline in conditions peaking some 1500–2000 years BP (§6.3.2). A temporary amelioration in climate occurred 1000 years ago but this was followed by a significant deterioration – the Little Ice Age – from which the planet is currently recovering (§6.4).

6.3.1 The climatic optimum

Global temperatures were perhaps 1°C warmer during the Climatic Optimum than today, with significant regional variation (Fig. 6.19). The origin of this warming was probably astronomical – the mid-latitudes of the Northern Hemisphere received 8% more radiation during summer at 9000 years BP than today – and there was little large-scale variation in the ocean circulation and surface properties (cf. deep water production signature in Fig. 6.17). However, the enhanced summer heating over Eurasia induced a 50% stronger Monsoon than now. This is reflected in enhanced upwelling in the Arabian Sea (Fig. 6.20), and hence cooler surface temperatures by

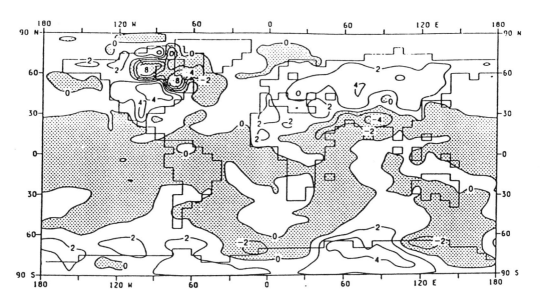

Fig. 6.19. Surface air
temperature at 9000 years
BP for Northern
Hemisphere summer,
relative to today, from an
atmospheric model
simulation of the period.
Areas that were colder at
that time are stippled; note
the remnants of the
Laurentide ice sheet over
NE North America.
[Reprinted, with kind
permission, from Fig. 16 of
Mitchell *et al.*, *J. Geophys.
Res.*, **93**, 16097–114 (1988).
Copyright of the American
Geophysical Union.]

1°–2°C in this part of the Indian Ocean. Minor accentuation of the driving Monsoon land–sea temperature contrast would have resulted.

One important oceanic change during this period was a freshening of the surface layer of the eastern Mediterranean from 9000 to 7000 years BP. Evidence for this is seen in the stratigraphy of the basin's sediments through the intermittent intrusion of organic-rich muds, known as sapropels (Fig. 6.21). These are believed to derive from accumulation in the sediment of organic detritus from the upper ocean during periods of oxygen deprivation at depth. The latter can occur when freshening, hence lightening, of the surface waters prevents convection and ventilation of the deep and intermediate waters. A likely climatic cause is enhancement in the Indian and African Monsoons, leading to greater rainfall in the highlands of east Africa. Rainfall over central Africa was indeed higher at this time (Fig. 6.22), leading to increased flow in the River Nile. Sea level was also approaching modern values (Fig. 6.15); contemporary joining of the Black Sea and Mediterranean may have resulted in a major fresh-water influx. The cessation of deep mixing in the eastern Mediterranean ultimately affects the properties of the water exchanged with the Atlantic proper. At the LGM probably only western basin water mixed deep to provide the overflow (§6.2.2); during the Climatic Optimum it has been hypothesised that the water balance of the whole Mediterranean was sufficiently positive for the present Gibraltan exchange to be reversed. The cut in supply of warm salty water at depth, which today helps to de-stabilise the NE Atlantic and pre-condition its waters for over-turning (§1.3.2), may have reduced the strength of the thermohaline circulation. There is a hint of such a lessening in the $\delta^{13}C$ record (Fig. 6.18), although a likely poleward expansion of the Atlantic sub-tropical anticyclone, and thus of the North Atlantic Drift, would have mitigated the Mediterranean's effect by directing saltier surface water northward.

Fig. 6.20. Percentage abundance of *Globigerina bulloides*, a near-surface dwelling foraminifer, from a core in the Arabian Sea, over the last 100 000 years. Stronger summer Monsoons cause more upwelling and hence the colder temperatures that this species prefers; hence an increase in abundance reflects stronger Monsoons. Using data from Clemens *et al.* (1991).

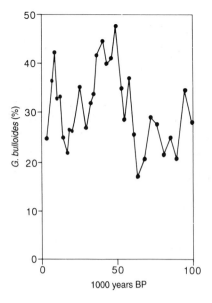

Fig. 6.21. Comparison of fresh-water events, when sapropels are laid down, in the Mediterranean over the past 350 000 years with the open ocean $\delta^{18}O$ record. Sapropel events can occur during glaciations (low $\delta^{18}O$) or interglacials (high $\delta^{18}O$), although they are more common during interglacials. [Fig. 1 of Thunell *et al.* (1983). Reproduced by permission of the Cushman Foundation.]

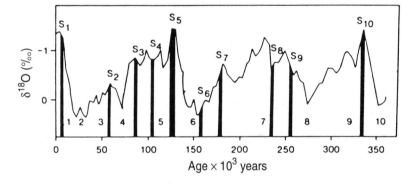

6.3.2   The last 5000 years

The climate since the end of the Younger Dryas has been dominated by decadal to century scale variability. Potential causes for this high frequency fluctuation will be discussed in later sections in this chapter, and in Chapter 7. However, longer term trends, such as the Climatic Optimum of the last section, can still be discerned. One such period – of generally cooler and wetter climate – occurred from roughly 4500 years BP to about 1200 years BP (c. 800 AD), with a period of warmer climate around 2000 years BP. This has sometimes been called the Iron Age cool epoch. Recorded history is now entered and tempting links between cool climates and unsettled periods in the history of early civilizations (Egypt, Mesopotamia and Greece – 1200–700 BC and the European Dark Ages – 300–800 AD) can be made, although possibly inaccurately.

An oceanic role in climatic deterioration at these times is not well established. There is some evidence for lessening of deep water formation in the North Atlantic coinciding with the two cool periods in this epoch (see

Fig. 6.22. Lake levels over Africa and the Near East during the Climatic Optimum. High levels shown by black dots, low levels by open circles and levels similar to today by open triangles. Note the prevalence of high levels over Saharan and Central Africa. [Adapted from Fig. 14 of *Lake levels and climate reconstruction*, by Street-Perrott and Harrison (Copyright 1985), in *Paleoclimate analysis and modelling*, edited by A. D. Hecht. Adapted by permission of John Wiley & Sons, Inc.]

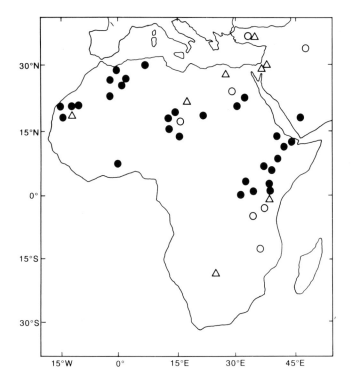

Figs. 6.17 and 6.18), although whether this is cause or effect is unknown. Greater sub-polar rainfall producing enhanced run-off into the northern North Atlantic, or an increased imbalance in the transport of atmospheric water vapour from Atlantic to Pacific leading to greater return, oceanic, flow through the Arctic could give this result. However, many other reasons for such climatic change could be invoked, as will be seen in §7.2.4.

For several hundred years around 1000 AD the climate warmed (Fig. 6.23). Many alpine glaciers retreated. From the evidence of Scandinavian exploration and settlement of the northern Atlantic it is likely that the North Atlantic Drift penetrated at least as far north as today, if not further. However, warming was not globally synchronous. For example, the east Pacific and Antarctica appear to have been warmer 200–300 years earlier.

From about 1300 until early in the twentieth century the global climate cooled again. In some regions during this period mountain glaciers expanded considerably giving rise to the term 'Little Ice Age'. As Fig. 6.23 suggests this period was not uniformly cooler, nor were changes in climate coincident around the globe. The main differences from the present climate, however, appear to have been increased storminess with accentuation of the zonal/meridional contrasts between the atmospheric circulation of the two hemispheres. The Southern Ocean seems to have experienced stronger westerlies, while enhanced Rossby wave dominance of the Northern Hemisphere circulation led to more meridional flow, and particularly cold northerly inflow over eastern North America and western Europe.

Despite the increase in historical references since 1300, and some

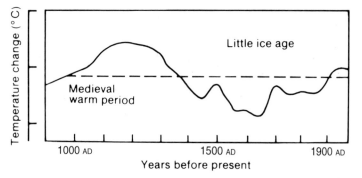

Fig. 6.23. Global surface air temperature over the past 1000 years. The scale is in °C departure from the 1950–80 mean. Note that the average global temperature over this millenium is lower than this latter mean. [Fig. 7.1c of Houghton *et al.* (1990). Reproduced with kind permission of Cambridge University Press.]

instrumental records, relatively little is known about the changes to the marine climate. There is some evidence for cooling over the eastern North Atlantic by up to 1.5°C, with perhaps a compensating increase in the central Atlantic. A similar pattern of stronger winds and lower sea surface temperatures is suggested in sediment records of the east Pacific. The main Pacific climate signature – El Niño – tends to show a reduced frequency during the seventeenth and eighteenth centuries (Fig. 6.24), although such a trend is slight and may be due to lack of evidence.

6.4 Marine climate change during the twentieth century

In 1854 Lieutenant Maury of the US Navy instigated the first scheme to encourage all ships, both merchant and naval, to regularly measure, and report, sea surface temperature. This idea was soon adopted by the British Admiralty, through the intervention of Admiral Fitzroy. Since that date this concept of 'Ship of Opportunity' measurements of an increasing set of oceanographic and meteorological variables has acquired international standing, and standards of instrumentation. An ever-growing data bank of ship-board observations now exists with which to examine marine climate over the past 140 years.

The data set is, nonetheless, imperfect. The number of participating vessels fluctuates over time; the general increase of the first few decades has been replaced by more variability due to economic cycles and fluctuation in ship tonnage. Merchant ships tend to follow well-used routes between trading centres leaving large areas of the world ocean unsampled (Fig. 6.25). The southeast Pacific, much of the Southern Ocean and the Arctic have very few observations for the entire period since 1854. Shipping routes also evolve with time, as trade between nations fluctuates, so that one region of the ocean can be well sampled for decades but now sees fewer ships. Periods of global warfare tend to drastically curtail measurement in some areas but massively expand it in regions of active naval warfare such as the central and west Pacific during the mid 1940s. A further difficulty with the data set is standardising the instrumentation (§6.4.1). This is now done routinely but evolving standards, and the earlier tendency for different countries to require their own standards can produce inconsistencies in the record.

Despite these problems much can be inferred from the observational record. The global sea surface temperature trend of the last century (Fig.

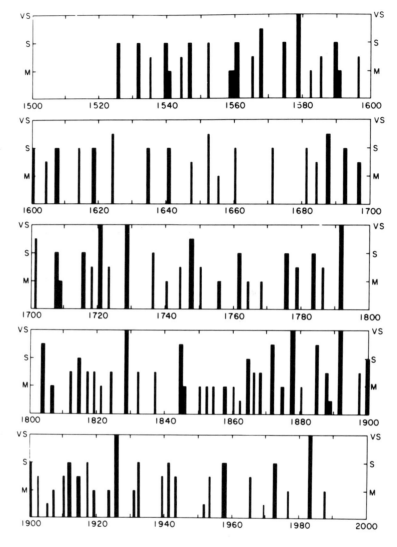

Fig. 6.24. Record of El
Niño events, both timing
and strength, over
1525–1987. Note the
enhanced frequency during
the nineteenth and
twentieth centuries,
compared to the
seventeenth and eighteenth.
[Fig. 32.1 of Quinn and
Neal (1993) in *Climate
since AD 1500*, edited by R.
S. Bradley and P. D. Jones.
Reproduced with
permission of Routledge.]

6.26) matches that over land (Fig. 1.29) well. Clear links have been
established between marine and terrestrial climate change, such as over the
Sahel (§5.1.3). The global changes in marine climate this century will be
examined in §6.4.2 and the Atlantic, Pacific and Indian Oceans dealt with in
turn in §§6.4.3–6.4.5.

6.4.1   *The instrumental record*

Imagine the contrast in observing conditions for a ship's officer, fulfilling
the routine measurement chore, between a sub-tropical evening and a
stormy sub-polar winter's night. In the first case it would be pleasant to be
on deck in a cooling breeze; one would feel at leisure and take time over the
observation. In the second situation, the rolling deck, gale-force winds,
cascading water from huge seas would make observation a trial to be

Fig. 6.25. The growth, and route-bias of sea surface temperature measurements over the oceans. The top diagram shows the network of observations in January of 1875, the bottom diagram of January 1975. Ship tracks are clearly seen in both months and even in 1975 little data is available south of 40°S. [Fig. 6 of Oort (1988). Reproduced with kind permission of Kluwer Academic Publishers.]

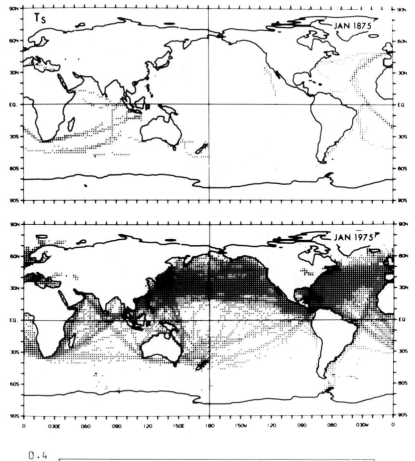

Fig. 6.26. Variation of global temperature since 1860, relative to the 1951–80 mean. Two estimates of global sea surface temperature (dashed and solid lines), night marine air temperature (crosses) and land air temperature (dots) are shown. While there is some discrepancy between these various measures of temperature the major trends are evident in all sets. [Fig. 7.9 of Houghton *et al.* (1990). Reproduced with permission of Cambridge University Press.]

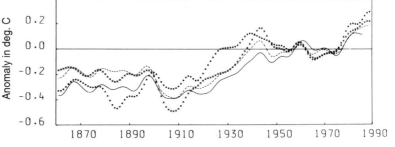

finished as rapidly as possible, if attempted at all. Understandable human error and weather-led bias in the places and times of observation are only one of the problems a user of the marine climate record faces. Instrumentation has changed dramatically since the heyday of sailing ships. How can data from different periods be compared?

Sea surface temperature has the longest, most geographically complete record of all marine meteorological variables. There was a history of measurement before 1854 which led to standard measurement from water collected in a canvas bucket dipped over the side. However, different nations employed standard buckets of varying size and degree of insulation, it later being recognized that evaporative cooling from the sides of a plain canvas bucket could reduce the measured water temperature. Heating or cooling from the air, or the ship's deck, could also be significant. The length of time between collecting the water sample and measuring its temperature also differed between nations. Some instructions suggested immediate measurement, to minimize evaporative cooling; the British recommended leaving 5 minutes to allow the water in the bucket to have mixed and attained a uniform temperature throughout.

Engine intake temperatures have been the more common form of sea surface temperature measurement since 1941. This has led to more systematic observations, although as water will be drawn in from slightly beneath the surface such measurements assume that the upper mixed layer is indeed well-mixed. Those observations still taken with the modern well-insulated bucket have a mean (cool) bias of less than 0.1°C. The pre-1941 measurements, however, have a greater bias. This amounts to approximately 0.3°C. Correction of these earlier temperatures by use of a physical model of evaporative cooling for specified buckets decreases the exaggerated seasonal cycle from summer to winter found from a uniform correction.

Systematic collection of marine air temperature measurements began at the same time as sea surface temperature. The difficulty of measurement, particularly in ensuring that radiation biases are not present, means that commonly only night-time observations are used for climate analysis. Even these, prior to 1894 and during World War II, suffered from biases due to inadequate screening from ship-board heat sources or poor ventilation. Such early measurements have been adjusted through use of the good (if later) correlation with sea surface temperature. The increasing size, and therefore deck height, of ships has also meant that correction for altitude is required of recent data.

Wind observations also have a long history and were regularly compiled from the late nineteenth century. Until the 1960s the regular method of measuring wind strength was the use of the marine Beaufort scale. This relates sea state, or the type of surface waves, to average wind speed in the lower planetary boundary layer. An experienced mariner would quickly be able to classify the sea state. This scale was devised in the early nineteenth century, but the classes were adjusted in 1946, it being realised that the old scale under-estimated wind speeds at the standard observing height of 10 m by about 1 ms^{-1}. The slow introduction of the new Beaufort scale to the world's shipping leads to a potentially false trend towards increasing wind speeds since 1946. Evidence for such false trends has been found in a few regions, notably the South China Sea. However, as will be seen in later sections the general increase in sub-tropical windiness in marine observations since World War II can be verified by comparison with land observations and variation in other related variables such as atmospheric pressure.

Since 1964 anemometers have become the standard instrument for wind speed on some, particularly American, vessels. These instruments are more accurate, but being inherently spot measurements are subject to biases due to their location on board ship, and their height above both deck and sea level. The Beaufort Scale remains important, however, for many vessels. The problems of compatibility between instruments and their exposure means that verification of wind changes by other means is particularly vital.

Cloudiness and humidity are also routinely measured. Cloudiness is historically a purely visual estimate and such observations are subject to pecularities in the observer. During the 1950s a change in reporting practice, from noting oktas, or eighths of the sky covered, to tenths, may have led to spurious trends towards increasing cloud cover. However, with the advent of geostationary satellites space-based observations of global cloud cover have become feasible. A data set for the 1980s is now available and in future space-based observation will supplement the more traditional measurements.

Humidity is calculated from the difference between the air and wet bulb temperatures. While salt contamination of the wick supplying moisture to the wet bulb can affect results, the main error here is probably associated with uncertainty in the background air temperature which has already been discussed.

The remaining commonly observed marine meteorological variable is atmospheric pressure. This also has records back to the nineteenth century and relatively good coverage since 1946. Instrumental design – either standard mercury or aneroid barometers – has been accurate to less than 1 mb throughout this time. Barometers are often on the ship's bridge and so measurement is not so subject to the vagaries of the weather. However, several problems exist with this data set. Readings from mercury barometers need adjustment made for temperature, because thermal expansion or contraction of the liquid metal can be significant. Opening or closing of cabin doors, or draughts in strong wind conditions, can cause significant changes to the pressure field within the ship, relative to that outdoors. The height of the instrument above sea level is also crucial. Atmospheric pressure decreases by about 1 mb for every 10 m of altitude near sea level so the difference in deck height between the laden and empty states of the same vessel will introduce error. The typical magnitude of each of these potential errors is 1–2 mb. However, as these errors tend to be random through a fleet of ships a sufficiently large number of observations over a month in any one area will yield a representative monthly average.

Rainfall is an important climatic variable but is essentially impossible to measure on board the moving, wave-splashed platform of a ship or buoy. However, there is a reasonable correlation between the outgoing long-wave radiation (OLR) measured from satellite and rainfall, particularly in the tropics. This is because the temperature, and so height, of a cloud top – the radiating level – is linked to the strength of convection and so amount of rainfall. Satellite measurements now provide a means of estimating oceanic rainfall that extends back to 1974.

Measurements of surface salinity and sub-surface temperature and salinity have much less extensive records than the above variables. Most of these measurements come from research vessels and so almost exclusively post-date 1930. They are generally much more accurately measured than most meteorological variables. Since the 1970s, additional sub-surface temperature, and more recently salinity, measurements have become available through deployment on merchant ships of expendable bathythermographs (XBTs) and their successors. These instruments, less accurate than their research vessel cousins but still reliable to within tenths of a degree Celsius, are regularly supplied as part of major research programmes such as TOGA. They are adding greatly to knowledge of the structure of the upper 500 m of the ocean.

An instrument under design which could revolutionize sub-surface oceanographic measurement is the auto-sub. This will take measurements in an unmanned, automatic, mode along pre-programmed routes, moving up and down in the water column and transmitting observations by satellite link to a land station when it is at the surface. While deployment of this instrument is still in the future, if it can be produced cheaply in large quantities regular oceanic monitoring will become feasible.

6.4.2   Global trends in marine climate

The global sea surface and night-time marine air temperature records (Fig. 6.26) show essentially the same trends as the terrestrial air temperature since 1860. Each contains a strong increasing trend from 1920–40 and again from 1975–90. However, during each period the land warmed faster than the sea, leading to a global land–sea temperature contrast roughly 0.2°C greater than average[3]. The heat storing capacity of the ocean is evident during the static period of 1950–70 when the land–sea difference became almost 0.1°C weaker than average. The net increase in temperature since the late nineteenth century is approximately 0.4°C. In Chapter 7 the potential causes for this trend will be discussed.

Geographically, the net trend is seen over the globe but components of it can vary significantly. For instance, the cooling of the 1950s and 1960s largely took place over the Northern Hemisphere. The Northern Hemisphere generally shows greater variability, because of the predominance of land and the less zonal atmospheric circulation. More recent cooling over the North Atlantic and Barents Sea is in marked contrast to the general warming which is notably uniform over the mid-latitudes of the Southern Hemisphere (see Fig. 7.24).

Other global marine climate parameters are much less well known. There appears to be a strengthening of oceanic wind speeds, concentrated in regions of the tropics and North Atlantic, but the problems with this variable discussed in §6.4.1 must be remembered. Supporting evidence exists from increasing wave heights in the North Atlantic, coastal wind measurements and land and sea pressure changes. These will be discussed in the appropriate ocean basin section.

[3] Most anomalies in this book are relative to the 1951–80 period, unless otherwise stated. In the mid-1990s the succeeding 30 year period, 1961–90, will begin to be the base for many such comparisons. However, the relatively static climate of the currently used period provides a more constant baseline than one overlapping the rapidly changing 1980s.

Fig. 6.27. Anomalies in
sea-ice extent in (a)
Northern and (b) Southern
Hemispheres over 1970–90.
[Fig. 7.20 of Houghton *et
al.* (1990). Reproduced with
permission of Cambridge
University Press.]

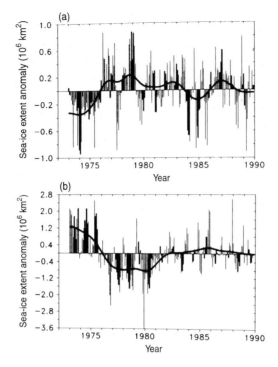

The numbers of tropical hurricanes have also increased in several regions (North Atlantic, Australian region, eastern North Pacific and SW Indian Ocean). This could be due to increased tropical sea surface temperatures (SST) providing more latent heat energy to the troposphere of hurricane-forming areas. However, some of the largest SST increases have occurred in the Bay of Bengal (§6.4.5) where no long-term increase is observed. It is likely that more hurricanes are now discovered from aircraft, ships and satellite than earlier in this century, so this increase must be treated with caution.

A climatology of oceanic cloudiness exists, stretching back to 1930. It suggests that mean cloudiness in the Northern Hemisphere has increased by 3–4%, and in the Southern Hemisphere by 1–2%. A particularly marked increase was recorded during the 1940s and 1950s. However, the Ocean Weather Ship network – a collection of vessels which remained fixed in location purely to provide weather observations – shows no evidence for such changes, even where passing merchant ship data does so. While merchant ship data generally agree with land-based changes in cloudiness there remains doubt about the reliability of inferences from this data.

Of major importance for shipping, and for detection of potential greenhouse warming, is the abundance and thickness of sea-ice (see §7.3.2). Global satellite coverage exists since 1974. In both hemispheres this shows significant variability about the mean (Fig. 6.27) but no trend. Analysis of earlier sea-ice records from ships and harbour authorities back to the 1950s supports this general finding. There is evidence for decadal scale fluctuation in the Atlantic but no long-term trend. Sea-ice thickness can only practicably be measured by upward-looking sonar from submarines. Extensive data occurs only in the Arctic where some evidence of a 15% decrease in ice thickness between 1976 and 1987 north of Greenland has been found. As no continuous data set exists this could just be a reflection of the amplitude of interannual variability.

The description of trends in the oceanic circulation and density field is severely hampered by the paucity of data (§6.4.1). Evidence of change in individual ocean basins will be documented where appropriate. However, here it should be noted that intermediate and deep waters in the Atlantic have warmed by 0.1–0.2°C since 1957, which will eventually have an effect on the strength of the global thermohaline circulation.

6.4.3   *Marine climate change over the Pacific Ocean*

The dominant climatic signal in the Pacific on sub-century scale is the El Niño/Southern Oscillation (§5.2). The frequency of the warm events in this cycle varies strongly over the last century (Fig. 6.24). The 1930s were a period of few events, followed by a prolonged warming from 1939–42. As a further warm event followed in 1944 the entirety of the early 1940s saw anomalously warm conditions over the tropical Pacific. Since then El Niño events have occurred with some regularity, apart from an absence during the period 1976–81. A further prolonged warm event occurred during 1991–4. Both prolonged El Niños this century have the character of an initial 'normal' event followed by a further year or more of anomalous heating and convection in the central Pacific. The longevity of the 1940s event was linked to the persistence of a very warm pool of water in the NW tropical Pacific, and reinforcement by mid-latitude Rossby waves (§5.2.3). In contrast, the west Pacific was anomalously cool during the middle of the 1991–4 event and the event was sustained by continuing convection over the west central Pacific (Fig. 6.28).

Underlying the variability imposed by ENSO are some distinct decadal-scale changes in the atmospheric and oceanic circulation of the Pacific Basin. The Trade winds of the tropical Pacific have strengthened over the last 40 years by about 1 ms^{-1} in the east Pacific, the central North Pacific, near the Hawaian Islands and the approaches to the Indo-Australian Convergence Zone in the west Pacific (Fig. 6.29). These changes are consistent with surface pressure trends, and in the east Pacific with coastal sea level variation. They appear to be linked to a stronger Walker circulation and a steepening of the meridional pressure gradient over the tropical North Pacific. Further north, an abrupt shift in the strength of the Aleutian Low in the mid-1970s has led to more intense storms in the North Pacific, and more poleward penetration of warm tropical air over Alaska while polar air spread further south in the central North Pacific.

The latter changes are visible in the 1981–90 global temperature anomaly field (Fig. 7.24). However, prior to this decade the atmospheric circulation adjustments had only minor impact on the Pacific sea surface temperature (Fig. 6.30). There is a significant warming in the west Pacific coupled to the heightened atmospheric convergence into this region. There is also some

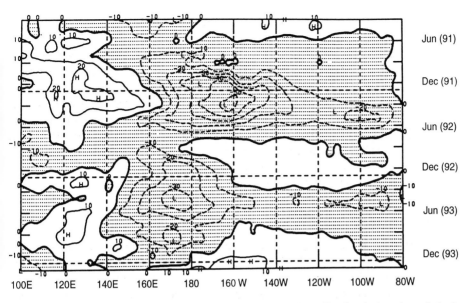

Fig. 6.28. Anomalous outgoing long-wave radiation (OLR) in the equatorial Pacific. Time is along the vertical axis and longitude along the horizontal. Negative anomalies (i.e. low temperature or high cloud tops) are shown by stippling. Anomalies are computed relative to 1979–88. Note the retention of anomalously high convection over the central Pacific in late 1992. [From the *Climate Diagnostics Bulletin* (1994b). Reproduced with permission of the Climate Analysis Center, USA.]

Fig. 6.29. Surface wind trends over the period 1950–81. The vectors are only plotted if they are significant at the 95% level. [Fig. 15 of Cooper *et al.*, *Int. J. Climatol.*, **9**, 221–42 (1989). Reproduced with kind permission of the Royal Meteorological Society.]

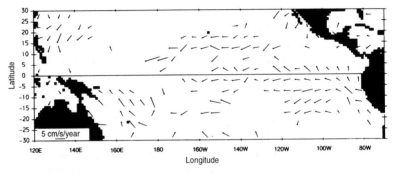

Fig. 6.30. Sea surface temperature trends over the period 1950–81. Barred regions are significant at the 95% level, horizontal bars implying decreasing temperature and vertical bars increasing temperature. Contour interval is 0.5°C, with the solid contour representing +0.25°C and the dashed contour −0.25°C. [Fig. 16 of Cooper *et al.*, *Int. J. Climatol.*, **9**, 221–42 (1989). Reproduced with kind permission of the Royal Meteorological Society.]

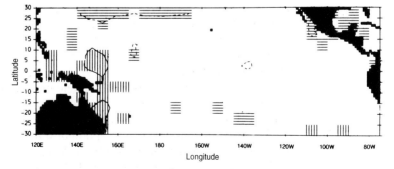

cooling over the central North Pacific, part of a more extensive cooling which intensified over 1950–90 (Fig. 7.24). This is presumably related to a longer term strengthening of the Aleutian Low that was merely accentuated during the 1970s.

Fig. 6.31. Normalised index
of atmospheric mean sea
level pressure difference
between Ponta Delgada in
the Azores and
Stykkisholmur, Iceland
(solid line), representing,
through geostrophy, an
index of westerly wind
strength. The surface
temperature over the 120
year period for the region
45–60°N, 5°W–35°E is also
shown, by a dashed line.
[Fig. 7.22 of Houghton *et
al.* (1990). Reproduced with
permission of Cambridge
University Press.]

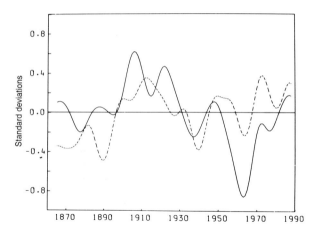

6.4.4 Marine climate change over the Indian Ocean

The Indian Ocean is dominated by the seasonal oscillation between NE and
SW Monsoons (§2.10.4). While this circulation is moderated by the ENSO
cycle of the Pacific it obeys its own dynamics (§5.2.6). A general strengthening of both Monsoons since 1946 is suggested by the marine wind data. This
can be verified by comparison with land and sea pressure trends, at least for
the NE Monsoon season, in the Mozambique Channel. However, there is
no clear trend in SW Monsoon rainfall over India (Fig. 5.39) as support.
Nonetheless, a clear increase in sea surface and air temperature over much
of the Indian ocean north of 30°S is found, particularly in the 1980s (Fig.
7.24) but also earlier. This has, however, not led to any increase in
hurricanes (see §6.4.2).

6.4.5 Marine climate change over the Atlantic Ocean

The Atlantic Ocean is the best observed ocean, and because of the strong
cyclogenesis associated with the Gulf Stream and the formation of a
significant proportion of the world's deep water it is the most climatically
sensitive ocean. Several major decadal-scale fluctuations in the climate of
the Atlantic have been experienced this century. A major intensification of
the prevailing westerlies in the mid-latitude North Atlantic occurred during
the first third of the twentieth century (Fig. 6.31). Such a flow arises from
increased cyclogenesis in the North Atlantic, leading to higher winter, but
lower summer, temperatures in western Europe. The notable global
warming observed during the 1920s and 1930s (Fig. 6.26) was strongest over
the North Atlantic and Arctic, consistent with greater transfer of heat from
ocean to atmosphere.

Since the 1940s the Atlantic has seen two further significant climate
anomalies. During the 1950s to mid-1970s a strong reduction in westerly
flow in the northern mid-latitudes led to cooling of western Europe. This
was associated with a decrease in the temperature of the northern North
Atlantic and a weakening of the pressure gradient between the Icelandic
Low and Azores High (Fig. 6.31), both decreasing in strength. A similar

Fig. 6.32. Changes in mid-ocean temperature profiles in the North Atlantic (dashed line) and North Pacific (solid line) between 1957 and 1981. [Fig. 7.18 of Houghton *et al.* (1990). Reproduced with permission of Cambridge University Press.]

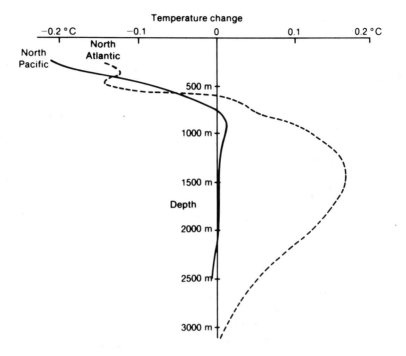

cooling of the mid-latitudes of the South Atlantic also occurred during this time. However, the limited data from this period in the South Atlantic suggests that warming occurred in the sub-tropics. This latter pattern has continued and, combined with cooling in the sub-tropics of the North Atlantic, has led to a southward displacement, and weakening, of the ITCZ over west Africa (§5.1.3). The strength of this anomaly in the northern summer has probably led to the drastic decline in Sahel rainfall since 1960 (Fig. 5.13).

These changes have also been associated with a strengthening of the Trade winds in both North and South Atlantic (Fig. 5.17), validated by comparison with land based pressure and wind observations. Such changes appear to be driven by equatorward and eastward movement of the sub-tropical Atlantic Highs, particularly during the northern winter.

Significant changes in the ocean circulation have also been detected during this period. About 1968 a fresh, cold water mass possibly originating in the Canadian Arctic several years earlier joined the circulation of the North Atlantic sub-polar gyre – this has become known as the 'Great Salinity Anomaly'. Its passage around the gyre was tracked for about a decade, and the anomaly may have weakened the rate of deep water formation. It has been hypothesised that this may be part of a coupling between ocean and atmosphere, with a timescale of several decades. The salinity anomaly would have stemmed from enhanced snowfall over Canada, and subsequent injection of extra melt water into the Arctic. The arrival of the anomaly in the North Atlantic would help to suppress cyclogenesis by lowering the North Atlantic temperature and weakening the northward heat flux in the Gulf Stream system by inhibiting deep water formation. This, in turn, would produce a colder northern hemisphere circulation and turn off the fresh water source over Canada. There is evidence that the magnitude of the Gulf Stream decreased during the time of the Great Salinity Anomaly. The surface waters cooled (Fig. 6.32) while intermediate waters warmed, lowering the density contrast supporting the current. It must be stressed, however, that such a mechanism is highly speculative.

Over the past two decades this cooling of the northern Atlantic and warming of the sub-tropical South Atlantic at all times of the year has strengthened (Fig. 7.24). However, it has been associated, in the North Atlantic, with a return to an enhanced westerly regime (Fig. 6.31), an increase in mean mid-ocean wave height, and slight warming over western Europe. This most recent period of dominant westerlies therefore has very different climatic characteristics to those experienced in the earlier part of this century. Chapter 7 will suggest possible reasons for this conundrum.

Further reading

A complete reference list is available at the end of the book but the following is a selection of the best books or articles to follow up particular topics within this chapter. Full details of each reference are to be found in the Bibliography.

Bigg (1992c): A short introduction to climates of the past.
Bradley and Jones (1993): This, plus the other articles in the same book give a comprehensive coverage of climate over the past 500 years. The concluding article, in particular, is a very readable summary.
Crowley and North (1991): An extremely readable, and well referenced, discussion of climates of past times.
Frakes *et al.* (1992): A good coverage of climatic change over the past 600 million years from a geological perspective.
Goudie (1990): A readable coverage of glacial and more recent climatic change.
Houghton *et al.* (1990): Careful discussion of the evidence for current and future climatic change.
Huggett (1991): An interesting, if sometimes eccentric, view of how the various parts of Earth have contributed to climatic change.
Lamb (1977): Now a somewhat old text but a very comprehensive coverage of climatic history.

7 *The ocean and climatic change*

Previous chapters have explored the varied ways in which the ocean and atmosphere interact, and the importance of this interaction to the Earth's present and past climate. The principal question in climatology today, however, is the path of future climatic change. In the 1970s the Northern Hemispheric cooling of the previous two decades, coupled with the future trend in solar forcing from orbital variation (§7.1.2), led to speculation that the Earth was entering the initial stages of the next glaciation. The global warming of the next decade, the 1980s, fuelled intense argument over the impact of anthropogenically-driven increases in the concentration of greenhouse gases. These debates are far from over, but they have served to greatly advance our understanding of the climate system. This chapter brings together the underlying causes of present climatic change, both natural (§7.1) and anthropogenic (§7.2), and identifies the ocean's contribution to the question 'where next?' (§7.3).

7.1 Natural variability

7.1.1 Solar variability

The Sun is the fundamental source of energy for the Earth's climate, yet remarkably little is known about the variability of its energy output. The theory of stellar evolution implies that the Sun's radiance has increased by 30% over the 3.5 billion years during which life has existed on this planet. It has been shown (§6.1) that a much stronger greenhouse effect than that seen today would have been needed to counter global glaciation early in the planet's history.

Shorter term fluctuation of the solar constant is only poorly known. There are currently only two sources of information: historical fluctuations in the number of sunspots and direct satellite observations of the last two decades. Sunspots are short-lived (a week or two) dark regions on the solar photosphere, caused by active convection within the Sun. Accompanying these cooler, dark spots are warmer, brighter regions, known as faculae. These typically cover a greater area of the Sun's surface so that periods of high sunspot numbers are times of higher net solar temperatures and therefore greater radiance (see Fig. 1.2). There is an approximately 11 year cycle in the number of sunspots (Fig. 7.1); satellite observations of radiance (Fig. 7.2) have shown that there was a 0.1% difference in the solar constant

Fig. 7.1. Annual mean
numbers of sun-spots from
1700 to 1991. [Fig. 11.1 of
Hartmann (1994), *Global
physical climatology*.
Reproduced with kind
permission of Academic
Press.]

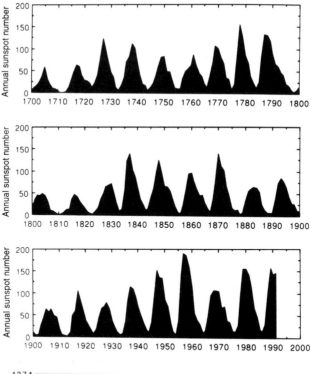

Fig. 7.2. Total solar
irradiance measurements
taken from the Solar
Maximum Mission
satellite, during 1978–89.
The units are Wm^{-2}; the
bar represents the
estimated absolute
accuracy of the measuring
instrument. Using data in
Hartmann (1994).

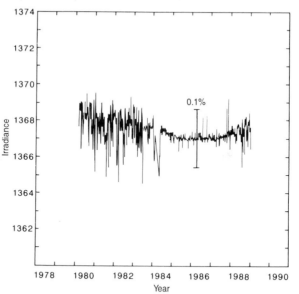

between the sunspot minimum in the mid-1980s and the previous maximum.

Considerable variation in the amplitude of the sunspot cycle, and
presumably therefore the solar constant's excursions, is seen in Fig. 7.1. A
roughly 80 year cycle has been hypothesised (e.g. one period of the cycle is

from peaks in amplitude in 1840–70 to 1940–60). There is also longer term variability: from 1645–1715 very few sunspots were observed, even during peaks in the 11 year cycle. This period is known as the Maunder Minimum and corresponds to one of the colder epochs of the Little Ice Age. It is believed that the solar constant was $0.25 \pm 0.1\%$ lower during this time. Similar long-lived minima have occurred over the past few thousand years at irregular intervals, perhaps twice a millenium. This has been established from the dependence of the strength of the solar flux of cosmic rays on sunspot number. Cosmic rays create atoms of the radioactive isotope of carbon, ^{14}C, during bombardment of atmospheric nitrogen (§3.7). Temporal variation of the ratio of $^{14}C:^{12}C$ is then recorded in the annual growth ring of trees, which can be reconstructed back almost 10 000 years.

The solar constant is thus the great unknown of climate. Current knowledge suggests that likely variation over the next century will be of limited climatic impact; the decrease during the Maunder Minimum represents only a quarter of the effective forcing already imposed by changes in greenhouse gas concentration over the last 200 years. However, the possibility of future changes in solar behaviour outside our experience cannot be ignored.

7.1.2 Orbital changes

Perturbations in the Earth's orbit have played a critical role in the glacial cycling of climate over the last million years (§6.2). These fluctuation continue (Fig. 6.8). The orbit is slowly evolving towards reduced summer solar insolation in both hemispheres (Fig. 7.3), as all three orbital parameters are entering phases with small amplitude. This should lead to global cooling, but the forecast reduction in insolation is only about a third of that which presaged the end of the last interglacial, 120 000–115 000 years BP. Even in the absence of anthropogenic effects a rapid glaciation over the next 10 000 years seems unlikely. Given the timescale of these insolation changes detectable climatic impact from orbital perturbation over the next 500 years will be neglected in the remainder of the chapter.

7.1.3 Volcanic impact on climate

Another driver of climatic change external to the climate system is the impact of volcanic eruptions. With essentially unpredictable frequency and severity volcanoes discharge quantities of dust, ash and trace gases into the atmosphere. The larger particulate matter can effectively reflect solar radiation, although even if they have been injected into the stratosphere such particles tend to be lost from the atmosphere by gravitational settling within at most a few months. However, a major gas emitted during volcanic eruptions is sulphur dioxide, SO_2. Sulphur dioxide is oxidised in the atmosphere to form sulphuric acid (equation (3.11)). The vapour pressure of this acidic gas is generally above saturation in the atmosphere and it readily condenses around existing aerosols or forms new sub-micron aerosols (§3.5.1). If the volcanic eruption is powerful enough to inject this gas into the

Fig. 7.3. Distribution of
departure of daily average
insolation from average
values over the period from
150 000 years ago until
20 000 years into the future;
(a) northern summer
solstice, (b) northern winter
solstice. Anomaly contours
are in Wm⁻². [Fig. 11.3 of
Hartmann (1994), *Global
physical climatology*.
Reproduced with kind
permission of Academic
Press.]

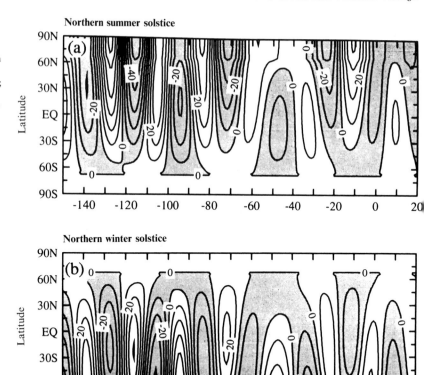

Fig. 7.3. Distribution of departure of daily average insolation from average values over the period from 150 000 years ago until 20 000 years into the future; (a) northern summer solstice, (b) northern winter solstice. Anomaly contours are in Wm^{-2}. [Fig. 11.3 of Hartmann (1994), *Global physical climatology*. Reproduced with kind permission of Academic Press.]

stratosphere the resulting sulphuric acid aerosols have a lifetime exceeding a year.

Several months after such an explosive eruption the aerosols will have formed and, if the eruption was in tropical latitudes, have provided a shield to solar radiation around the globe. Volcanic impacts on climate, however, are complicated by the particles' absorption of infra-red radiation. If the particles are predominantly below 1 μm in diameter the net climatic effect is a surface cooling, while particles above 2 μm tend to warm the troposphere. In general, sulphuric acid aerosols tend to be small so volcanic eruptions usually produce a net cooling. The 1991 eruption of Mt. Pinatubo (the Phillipines, 17°N) is thought to have cooled the globe by about 0.3°C during 1992–3 (Fig. 1.18). Similar short term coolings were associated with eruptions of Mt. Agung in 1963, Krakatoa in 1883 and Tambora in 1815.

The climatic impact of any volcanic eruption is unlikely to last more than three years. Given the timescales of processes within the ocean, long-term climatic change is therefore unlikely to be driven volcanically, unless there are large eruptions every few years. Volcanism is, nonetheless, irregular. The last century has experienced relatively few major eruptions while the acidity of ice-cores in Greenland – which is probably a strong function of stratospheric and tropospheric sulphur aerosols – indicates that volcanism

Fig. 7.4. Volcanic fall-out consists of acidic material so the record of acidity (in μEquivalents/litre) in ice-cores is a proxy for volcanic activity. The left picture shows the acidity of annual layers in ice from Crete, Greenland, the right, from Dome C in Antarctica. [Fig. 31.5 of Bradley and Jones (1993) in *Climate since AD 1500*, edited by R. S. Bradley and P. D. Jones. Reproduced with permission of Routledge.]

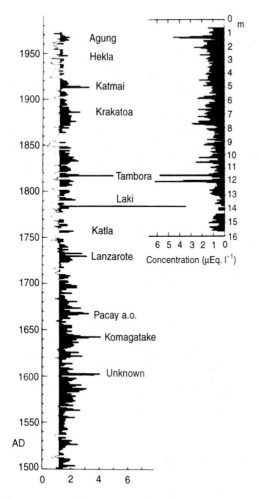

was more pronounced during the Little Ice Age (Fig. 7.4). It is possible that volcanic aerosols contributed to cooling during this period but unlikely that the cooling was driven by volcanism alone.

7.1.4 Cometary impact

The collision of the fragments of Comet Shoemaker–Levy 9 with Jupiter in July 1994 confirmed the potential for climatic change of a cataclysmic nature by cometary impact. Since the late 1980s evidence for a terrestrial collision occurring 65 million years BP has mounted. A layer of iridium, an element rare on Earth but revealed through emission spectra to be more abundant in comets, has been found in the geological record at the Cretaceous–Tertiary boundary at many places around the world. A crater near the Yucatan peninsula of Central America has been identified as the impact site of a 10 km diameter comet or asteroid around that time. Such an impact would hurl enormous amounts of dust into the stratosphere, possibly cooling the planet by tens of degrees for several months. An impact with the ocean would also add great quantities of water vapour to the atmosphere and create a wave perhaps 1 km high flooding much of the land area of the planet. The mass extinction of species, including the dinosaurs, at this time is likely to have been a direct result of such an impact.

Such catastrophes are fortunately rare. However, smaller impacts are more frequent. There are a number of geologically recent impact craters around the globe, the best known being Great Barringer, or Meteor, Crater in Arizona, some 30 000 years old. More recently, in 1908 a small asteroid landed in an uninhabited part of Siberia, demolishing forest over several hundred thousand square kilometres. The global climate in the early 1900s was anomalously cool (Figs. 1.26, 6.26) but it is noticeable that, globally, 1908 was 0.2°C colder than the years around 1905, perhaps reflecting an impact-driven increase in stratospheric aerosol. Extra-terrestrial collisions thus remain infrequent and unpredictable causes of short-term climatic change of potentially immense proportions.

7.1.5   *Internal climatic instability*

Much of the discussion in Chapter 6 invoked interaction between components of the climate system to explain climatic fluctuation. The post-glacial climate right up to the present decade has experienced oscillations in climate of decadal to century scale that are predominantly driven by feedbacks between the ocean, atmosphere, land and biosphere. This natural variability will make detection of anthropogenic effects difficult because their amplitude, as seen by fluctuations over the last 2000 years (Fig. 6.23), is commensurate with estimates of current greenhouse gas-induced warming.

Many of these internal couplings have already been discussed in Chapter 6, or will be amplified in §7.2. These include the up-to-decadal scale of the Southern Oscillation, fresh water forcing of various forms of the North Atlantic and the interdependence of storm tracks and sea surface temperature patterns. However, these are only a subset of the internal variability modes of the climate system. Numerical models of the atmosphere, coupled with the ocean, display oscillations in global temperature of decadal scale very reminiscent of observed temperature trends this century (Fig. 7.5). Models of the ocean alone, forced with random noise superimposed on the seasonal cycle of wind stress, heat and fresh water fluxes, display even longer oscillations. A density anomaly cycle of some 350 years has been induced in the Atlantic circulation, and fluctuations even on the 1000 year timescale have been produced in global ocean models.

Ocean models also exhibit an internal oscillation in the strength of the thermohaline circulation on decadal timescales due to pronounced local minima in $P-E$, the net water flux to the ocean. The observed minimum over the Greenland Sea produces such an oscillation in a numerical model of an idealised ocean basin. If the thermohaline circulation is weak, surface water passes slowly through this region and so gains in salinity. However, higher salinity means higher density water, and when the water is cooled in winter it will more easily convect, thus strengthening the thermohaline circulation. The consequent faster moving surface waters being drawn into

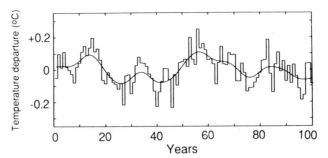

Fig. 7.5. The temporal variation of the deviation of global mean surface air temperature, in °C, of the Princeton coupled ocean–atmosphere model from its long-term average. [Fig. 6.2 of Houghton *et al.* (1990). Reproduced with permission of Cambridge University Press.]

the convection region will now be made less saline, slowing convective over-turning and starting the cycle once more. Similar changes in the properties of surface water in the North Atlantic will have an impact on the atmosphere through the magnitude and geographical location of latent heating.

Real or potential internal oscillations and feedbacks between seemingly disparate components of the climate system are constantly being discovered. This is an exciting field that will continually impinge on our discussion of potential anthropogenic impacts on climate.

7.2 Anthropogenic forcing of climate

The history of humanity's diaspora from Pliocene Africa extends over several million years, but unambiguous evidence of our impact on the Earth's landscape only post-dates the last glacial period. Removal of forest cover over western Europe, Asia and Australia due to farming or deliberate burning dates from about 10 000 BP. Anthropogenic modification of the Earth's atmosphere is of even more recent origin; such increases in atmospheric CO_2 and CH_4 date only from 1750.

The short time in which humanity has been able to influence components of the climate has, however, seen a drastic escalation in the Earth's population and so our ability to modify the environment which enfolds us. Our direct impact takes three forms: increase in various greenhouse gases through combustion and agricultural practices; changes in the aerosol number, type and distribution; and changes in the surface properties of the terrestrial environment. The evolution of these three mechanisms of change will first be discussed (§§7.2.1–7.2.3), followed by the resulting feedbacks that drive climatic change (§7.2.4), with particular emphasis on those oceanic.

7.2.1 Trace gases

Gases in the atmosphere that contribute to the greenhouse effect occur only in trace quantities but have a powerful warming effect on the tropospheric climate (§1.2.1). Apart from water vapour, which is highly variable in space and time because of its pivotal role in the hydrological cycle, these trace gases have atmospheric lifetimes sufficiently long for them to be relatively well-mixed through the global troposphere. Some have slightly higher

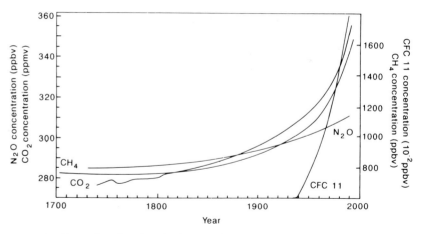

Fig. 7.6. Variation in the atmospheric concentrations of four greenhouse gases over the last 300 years. The data are a combination of (recent) direct measurement and ice-core records.

concentrations in the hemisphere of their principal production, but generally by a few per cent at most.

Despite this rapid mixing many greenhouse gases have significantly increased in concentration over the past 250 years, as measured in ice-cores or, more recently, direct observation (Fig. 7.6). Some of these gases, such as the chlorofluorocarbons, have only appeared in the atmosphere since their manufacture for industrial use. Others, such as CO_2 and CH_4, are showing change greater than has been observed in the recent geological past (Fig. 6.10).

Carbon dioxide is the most important contributor to the greenhouse effect among these well-mixed gases (Table 1.2). Being the product of combustion (or complete oxidation) of carbon-bearing compounds it is produced by burning of fossil fuels (6.0 ± 0.5 gigatonnes[1] of Carbon – GtC – per year as of 1990) and during deforestation (1.6 ± 1.0 GtCyr^{-1}). CO_2 has a natural cycle through the environment (§3.3), however, so much of this addition of long-stored carbon to the atmosphere is quickly exported to other components of the carbon cycle (see Fig. 3.4). It has been estimated that of the CO_2 of anthropogenic origin added to the atmosphere between 1850 and 1986 only $41 \pm 6\%$ has ultimately remained there. The increasing rate of addition has meant that this proportion has increased to $48 \pm 8\%$ for emissions between 1980 and 1989. Note that individual CO_2 molecules are not likely to have remained in one storage element of the carbon cycle; re-adjustment of equilibrium storages is occurring.

The ocean is a major store of carbon; some 2.0 ± 0.8 GtCyr^{-1} of the CO_2 added to the atmosphere are added to the ocean's biological and chemical reservoirs of carbon. Atmospheric CO_2 is increasing by 1.8 ppmyr^{-1}. As 1 ppm represents 2.12 GtC the current annual atmospheric storage of CO_2 is approximately 3.8 GtC. This leaves 1.8 ± 1.4 GtCyr^{-1} unaccounted for. Various mechanisms have been proposed for increased storage in the ocean. These include: a larger oceanic reservoir of dissolved organic carbon from more rapid cycling in the biological productivity cycle; greater flux of carbon between the hemispheres in the ocean; use of the roughly 0.3°C

[1] A gigatonne $= 10^9$ tonnes $= 10^{15}$ g.

Table 7.1. *Sources and sinks of methane (10^{12} grams/yr, after Houghton et al., 1992)*

Mechanism	Annual rate
Source	
Natural	
Wetlands	100–200
Termites	10–50
Ocean	5–20
Freshwater	1–25
CH_4 hydrate	0–5
Anthropogenic	
Coal mining and natural gas/petroleum industry	70–120
Rice Paddies	20–150
Enteric fermentation	65–100
Animal wastes	20–30
Domestic sewage	25
Landfills	20–70
Biomass burning	20–80
Sinks	
Atmospheric removal	420–520
Removal by soils	15–45
Atmospheric increase*	28–37

* Recent measurements suggest this may be decreasing.

cooler surface, or skin, temperature of the ocean for calculating air–sea CO_2 exchange rates (see equation 3.1); and greater biological consumption of carbon than previously assumed from the Redfield ratio (§4.3.3). Notwithstanding these contributions it is currently believed that an unknown terrestrial sink of CO_2 must exist in order to balance the added carbon budget. Some candidates for this sink exist: enhanced terrestrial productivity from atmospheric CO_2 increases nitrogen fertilization and re-afforestation in northern mid-latitudes. The size of such sinks are, however, not known.

Another major greenhouse gas, that has doubled in concentration over the last 250 years, is methane (Fig. 7.6). This gas has numerous sources (Table 7.1), mostly from biological decay occurring in anaerobic conditions (see §4.2.1), from venting of natural gas and burning of petroleum products. The major mechanism for removal of atmospheric CH_4 is via a complex reaction sequence with the hydroxyl radical. Minor removal mechanisms include soil uptake and photolysis in the stratosphere. Despite the very rapid increase in methane over the last century the latest atmospheric measurements suggest that the rate of increase is slowing equally rapidly. Equilibrium may be reached in the mid 1990s. It is not clear why this equilibrium is being established but it may involve a balance between expansion of rice paddies and destruction of natural wetlands.

Few of the various chlorofluorocarbon (CFC) compounds have natural sources and it is only since the 1930s that these chemicals have entered the

Fig. 7.7. Schematic
illustration of the
mechanism by which
chlorine is made available
for ozone destruction in
the spring stratosphere.
Paths (i) and (ii) indicate
the two ways ClO can be
broken down. See the text
for more details.

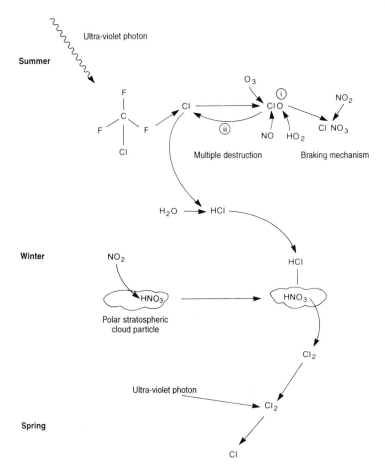

atmosphere (Fig. 7.6). These gases, however, have lifetimes of centuries as the only significant loss mechanism is photolytic destruction by ultra-violet light in the stratosphere. They are also very effective greenhouse gases because of the carbon–halogen bond. This has a major absorption band in the atmospheric water vapour window (see Fig. 1.9 and Table 1.2).

The destruction of these molecules in the stratosphere does not end their impact on the radiative properties of the atmosphere. The chlorine or bromine released on photolysis ultimately reacts with stratospheric ozone, converting it to oxygen molecules and chlorine monoxide (ClO). However, what has led to a major depletion in ozone in the polar stratosphere – the 'ozone hole' – is the ease with which ClO can react to release the chlorine atom for renewed attack on ozone (Fig. 7.7). This depletion is presently concentrated over Antarctica in the austral spring because wintertime reactions on ice crystals release chemically inert chlorine molecules (Cl_2), rather than atoms, into the atmosphere. These are only photolysed to the chlorine atoms that destroy ozone once sunlight returns to southern latitudes. In the early 1990s a similar depletion of spring stratospheric ozone has begun to be observed over the Arctic.

Fig. 7.8. Distribution of
total column ozone, in
Dobson units, over the
Southern Hemisphere in (a)
October 1992 (spring) and
(b) February 1993
(summer). Data from the
TOMS satellite, courtesy of
Rutherford Appleton
Laboratory.

V6 TOMS total O3 (Dobson Units), monthly average
Oct. 1992
Smoothed to 6 deg. latitude by 10 deg. longitude

Southern Hemisphere,
Orthographic Projection.

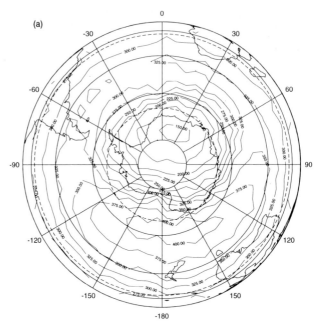

RAL GDF (User BIGG)   Plotted at 15:34:42 on 27-SEP-94

V6 TOMS total O3 (Dobson Units), monthly average
Feb. 1993
Smoothed to 6 deg. latitude by 10 deg. longitude

Southern Hemisphere,
Orthographic Projection.

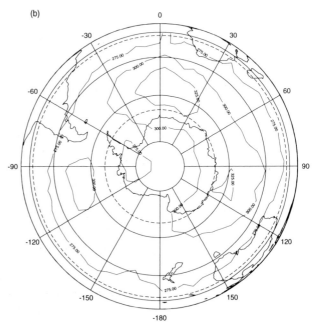

RAL GDF (User BIGG)   Plotted at 15:55:29 on 27-SEP-94

Away from the polar regions ClO tends to react with nitrogen dioxide to form a stable species, chlorine nitrate:

$$ClO + NO_2 \rightarrow ClONO_2 \tag{7.1}$$

This reaction acts as a brake on ozone destruction by limiting the repeated impact of any one Cl atom, in regions where ozone is also more easily formed by photolysis (equation (1.4)). However, if the atmosphere is cold enough ($< 200K$) for stratospheric cloud to form, as over Antarctica in winter, NO_2 is depleted by reaction with the clouds' ice crystals. Nitric acid is so formed, which can then react, again on ice crystal surfaces, with the hydrochloric acid (HCl) that is the main form by which Cl exists in the stratosphere, to release Cl_2 to the air. Thus in the wintertime Antarctic stratosphere there is not only less NO_2 to remove ClO but also production of Cl_2.

Depletion of stratospheric ozone is therefore concentrated over the Antarctic in spring (Fig. 7.8) while the strong zonal stratospheric winter flow around this continent persists, isolating the Antarctic stratosphere. Once this breaks down, in late spring, ozone-rich air from the tropics raises the ozone level. This annual cycle of depletion has been taking place since about 1980, and coupled with the slow destruction of stratospheric ozone elsewhere has meant that the global total mass of ozone has slowly been decreasing in recent years by 1–5% per decade. This globally slow, but regionally large, decrease in stratospheric ozone means less solar radiation is absorbed at height in the atmosphere. This accentuates any greenhouse warming (§7.2.4). The nature of the extra radiation now reaching the surface – ultra-violet wavelengths – may have an impact on the productivity of the biosphere causing a knock-on effect on the climate (§7.2.4). Measures have been taken internationally to limit emission of certain halocarbons; the implications of this for climate will be considered in §7.3.

The tropospheric concentration of a number of more minor greenhouse gases also show changes induced by humanity's activities. Nitric oxide (N_2O), produced in various biological processes (e.g. §4.2.2), shows a slow increase (Fig. 7.6). Tropospheric ozone, a by-product of photolytic reactions with industrial and automotive NO_x emissions[2], and CO have shown some evidence of increase. This may actually have implications for the increased tropospheric flux of ultra-violet light arising from stratospheric ozone depletion. The longer path length of light travelling through the troposphere, due to scattering, allows tropospheric ozone to absorb significant amounts of this extra flux. Sulphur dioxide, a major by-product of fossil fuel burning, has a short lifetime in the troposphere because of its rapid oxidation to sulphuric acid particles (equation (3.11)). Its climatic impact stems from the aerosols so formed.

7.2.2   Aerosols

Aerosols affect climate. They scatter and absorb radiation. They are critical for cloud formation and precipitation (§§2.8.2, 3.5, 4.4, 7.1.3; see §7.2.4 for a

---

[2] NO_x includes the various oxides of nitrogen, although not N_2O.

Fig. 7.9. Changes in annual
mean (a) surface air
temperature, in °C, and (b)
precipitation, in mm due to
deforestation of northern
South America, according
to an atmospheric model
simulation. [From Fig. 10
of Nobre *et al.* (1991), *J.
Climate*, **4**, pp. 957–88.
Reproduced with kind
permission of the American
Meteorological Society.]

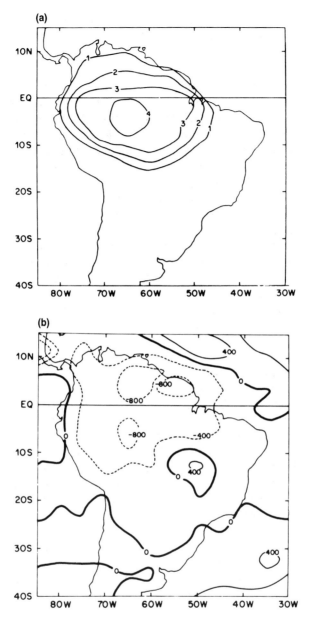

discussion of the relevant feedback processes). There are many natural
sources of aerosols which humanity has the capacity to alter but the two
major mechanisms for anthropogenically changing aerosol number and
size spectra are through industrial emission of SO_2 and the bio-marine
DMS flux to the air.

Sulphur is a significant constituent in many fossil fuels. Oxidation of this
sulphur upon combustion releases about 80 Tg (1 Tg = 10^{12} g) sulphur to
the atmosphere every year. Burning associated with deforestation releases
approximately another 7 Tg annually. Natural emissions of sulphur gases –
which include H_2S and DMS in addition to SO_2 (§4.3.5) – amount to only

about two thirds of this anthropogenic total. These sulphur gases readily oxidise to sulphuric acid, H_2SO_4. This is either adsorbed onto existing aerosols as an acidic and hydrophilic film or condenses to form particles in its own right.

Sulphur emissions are estimated to have doubled in the 40 years from 1940 to 1980. However, the atmospheric lifetime of sulphate aerosols is only days to weeks. Thus the stratosphere, and remote regions of the troposphere such as the South Pacific, have experienced no significant trends in aerosol concentration. The situation in the lower atmosphere over North America and Eurasia is very different. Some limited evidence of decreases in visibility and increases in cloud cover over these continents may be attributable to rises in aerosol concentrations. The most direct evidence of sulphate aerosol accumulation, however, is through the acidifying of rainfall over regions downwind of large urban and industrial areas. Greater abundance of sulphuric acid within aerosols has sometimes taken the pH of rainwater from its natural level of about 5.6 to as low as 2–3 during particular precipitation events[3].

7.2.3   Land surface changes

The last 250 years have seen dramatic changes to land use over much of the habitable globe. The predominant conversions of vegetation type – from forest to arable and savannah to desert – change the albedo of the land surface. More importantly, the moisture and thermal characteristics of the land change. The global climatic impact of desertification and deforestation, excluding any addition of greenhouse gases to the atmosphere through burning, is thought to be negligible. However, regional climates may change significantly from changing land use. For example, it is estimated that stripping the Amazon rainforest from northern South America could cut the regional rainfall by more than half and increase the temperature by 2°C (Fig. 7.9).

7.2.4   Climatic feedbacks

Direct changes to the environment resulting from humanity's activities are considerable, as shown in §§7.2.1–7.2.3. The climatic impacts of these changes, and particularly the oceanic contribution, involve numerous interactions within the climate system. Clouds and water vapour, and hence the oceans, lie at the hub of these feedbacks.

All direct anthropogenic changes to the environment alter the basic radiational properties of the planet. The raised levels of atmospheric greenhouse gases have little impact on the solar flux but ensure that more of the Earth's re-emitted infra-red radiation is re-cycled in the troposphere before the energy is finally lost. The mean energy in the climate system must

---

[3] pH is the common measure of hydrogen ion, H^+, concentration: pH = 3 is equivalent to $[H^+] = 10^{-3}$ mol m^{-3}. The lower the pH, the greater the concentration of free H^+ ions, or the more acidic the solution.

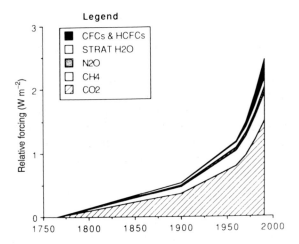

before the energy is finally lost. The mean energy in the climate system must remain unchanged, in the absence of variation in the solar constant or Earth's albedo (but see later), but more of this energy is now being cycled through the absorption, re-emission loop in the lower troposphere for longer (see Fig. 1.11). The net effect of this is to warm the troposphere and cool the stratosphere.

By 1990 this enhancement of the greenhouse effect decreased the longwave radiative flux leaving the troposphere (across the tropopause) by about 2.5 Wm^{-2}. This represents an accumulation of energy in the troposphere equivalent to about 1% of the absorbed solar radiation, which might be expected to have produced an increase in global tropospheric temperature of roughly 0.7°C compared to 1750.

The contributions of the various additions to the greenhouse gases over the last 250 years to this direct impact are shown in Fig. 7.10. Carbon dioxide has been the major contributor but it is noticeable that the low concentrations of chlorofluorocarbons, relative to CO_2, now contribute 10% of the total enhancement. The impact of a gas on the radiation spectrum is not merely a function of concentration, but crucially depends on where in the infra-red the gases absorb, and the depletion in emission over these bands by other gases already present. As CO_2 has historically been a major greenhouse gas, emission at some wavelengths absorbed by this gas was already significantly reduced prior to 1750. In contrast, the CFCs are strong absorbers in the emissively transparent atmospheric window (Table 1.2).

The radiational effect of ozone depletion in the stratosphere re-inforces that caused by enhanced greenhouse gas concentrations in the troposphere. Less stratospheric ozone means more solar radiation penetrates to the troposphere. This is short wavelength, ultra-violet radiation which is readily absorbed in the troposphere, the oceans and by terrestrial organisms. A cooler stratosphere and warmer troposphere will result.

The direct radiational impact of these changes to atmospheric gases leads to various indirect climatic effects. Warming the lower troposphere results in more evaporation from the sea and land surface as the air's capacity to

Fig. 7.11. The impact of
biological processes in the
upper ocean on the partial
pressure of CO$_2$ in the
upper ocean. The *-line
shows actual measurements
of *p*CO$_2$ in the Norwegian
Sea through a 12 month
period. The o-line shows
the concentration expected
from purely chemical
arguments related to the
temperature of the water
and the rate of chemical
dissolution of CO$_2$ in sea
water.

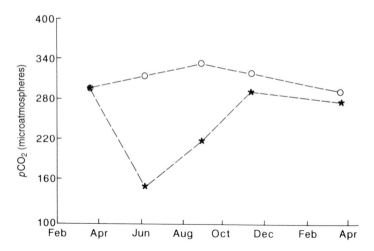

greenhouse gas (Table 1.2). A further enhancement of the greenhouse effect occurs in what is known as a positive feedback. This results in a net increase in trapped tropospheric energy of 1.6 times the non-water vapour enhancement.

Water vapour, however, results in cloud formation. Clouds cause a host of climatic feedbacks, some positive, some negative (Fig. 1.23), although the overall impact of cloud is to increase the Earth's albedo and so act to cool the planet. The proportion of sky covered by cloud – the cloud amount – has, however, potentially balancing feedbacks. High thin cloud such as cirrus is almost transparent to solar radiation but emits little infra-red radiation. Low thick cloud over the ocean tends to reflect solar radiation but has little impact on the outgoing infra-red radiation from the tropopause. An increase in global cloud cover by about 6%, with the present height distribution, would balance the current enhanced greenhouse effect.

A warmer troposphere has further implications for cloud processes. Higher concentrations of water vapour may lead to a higher liquid water content in clouds. This would give them greater reflective power and so provide a further negative feedback. This effect would be particularly strong for higher clouds, as low clouds already intercept most direct solar radiation. A warmer troposphere also changes the saturated adiabatic lapse rate (see Appendix C). Stability, and thus convective cloud formation, within the atmosphere, is controlled by the difference between the local lapse rate of the atmosphere – the environmental lapse rate – and the dry adiabatic lapse rate (DALR) within cloudless air, but the saturated adiabatic lapse rate (SALR) within cloud. An environmental lapse rate greater than the appropriate adiabatic lapse rate of a rising parcel of air promotes instability and convective cloud formation, while a lesser environmental lapse rate favours stability or suppresses vertical cloud growth. The SALR is less than the DALR because of the internal heating due to release of latent heat during condensation. If the troposphere warms then the amount of water vapour condensing will increase (see Fig. 2.5), causing the SALR to be even lower. This additional condensation enhances the tropospheric warming, but it also means that it may be easier for the troposphere to support convection. More, and thicker, cloud would result, reflecting more solar radiation and leading to a negative feedback. This, of course, relies on the mean environmental lapse rate remaining constant, but depending on how the warming is distributed in the troposphere the background lapse rate could be steepened or lessened. The former encourages convection, probably resulting in a negative feedback, while the latter discourages convection and gives a positive feedback.

The evolution of cloud properties over this century is relatively little known (§6.4.2). The true sign of the net feedback of clouds on climate in an enhanced greenhouse world is therefore not well established.

Cloud feedback is complicated further by anthropogenic modification of the natural aerosol population (§7.2.3). Sulphate aerosols tend to be sub-micron in size and thus lead to cloud droplets of similar diameter. Increases in sulphate aerosols lead to increases in cloud droplet concentrations. This decreases the mean radius of the cloud droplets. These two factors together lead to greater reflection of solar radiation and thus a negative feedback, which has been estimated to be currently as large as 1.3 Wm^{-2}, or half the addition of energy to the troposphere provided by the (radiative-only) enhanced greenhouse effect. There have also been anthropogenic additions to organic aerosols (§3.5.1) which are mostly smaller than 0.05 μm in diameter. These may enhance this negative feedback.

Sulphate aerosols also contribute a direct climatic negative feedback: scattering of solar radiation back to space. While this, and the indirect impact discussed above, are regional in nature (§7.2.3) it has been estimated that the loss of basic energy to the climate by this direct effect may contribute a further 0.5 Wm^{-2} of tropopause energy emission. The impact of sulphate aerosols on the climate may therefore go a substantial way towards negating any current enhancement of the greenhouse effect.

Sulphate aerosols are produced by natural sources as well as anthropogenically. The major such source is biological and marine, namely, the oxidation of DMS over the oceans (§4.4). Change to marine biological productivity is required in order to provoke change in this considerable source of marine aerosol. This may also be a consequence of the enhanced greenhouse effect, through several mechanisms. The most immediate is the increase in CO_2 available for photosynthesis in the ocean from the significant marine input of this gas as it fractionates between the atmosphere and ocean. From reaction (4.1) such increases will drive more vigorous photosynthesis. Greater biological production leads to more dissolved organic carbon in the mixed layer, and so more secondary production is possible because of the greater re-cycled nutrient base. Increased biological production, in turn, depletes the surface waters of CO_2, encouraging further sequestration of the gas in the ocean and slowing its atmospheric build-up: a negative climatic feedback (Fig. 7.11). Active production also deposits more carbon in the atmospherically inaccessible deep waters through sedimentation of organic detritus from the mixed layer. Bio-marine moderation of atmospheric CO_2 may be responsible for the 30% natural rise in CO_2 since the last glaciation (Fig. 6.10).

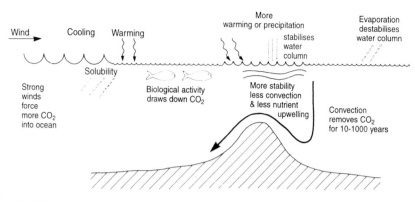

Fig. 7.12. Schematic illustration of mechanisms affecting absorption of CO_2 in the ocean.

Addition of nutrients to ocean waters from land effluent or greater wind-driven mixing can accentuate this negative feedback by stimulating production. Thus the observed increase in marine wind speeds over various regions in recent decades (§6.4.2) may indirectly slow the atmospheric CO_2 rise. There is little information on changes in productivity over the last century with which to validate such a hypothesis, although the absence of change in the transparency, and hence plankton mass, of surface waters in the North Pacific militates against such change. There is, moreover, an internal mechanism within the productivity cycle to moderate such an effect. Greater plankton populations lead to greater absorption of solar energy in the mixed layer and cooling by shading of the deeper water. This will increase the stability of the near-surface water and hinder wind-driven mixing, thus lowering the nutrient supply.

Marine productivity is also dependent on temperature and the general ocean circulation. The most vigorous production, and hence CO_2 draw-down, occurs in upwelling zones and regions of deep winter convection (cf. Figs 3.2 and 4.5). A heating of the surface ocean waters by the enhanced greenhouse effect would tend to stabilise the upper ocean (see Fig. 5.25) making mixing in these regions, and therefore the upward convective transport of nutrients, more difficult. This would allow a faster build-up of atmospheric CO_2 by slowing the ocean uptake, and lead to a positive feedback. In addition, warming the surface waters decreases the solubility of CO_2 (Fig. 3.1), also curbing ocean uptake. Warming of the ocean could also push the sub-tropical gyres further north; this would decrease the area of vigorous productivity and CO_2 draw-down in sub-polar waters. Recent climatic change, however, suggests a trend towards greater meridional temperature gradient, with increased westerlies, rather than a long-term northward shift of the Gulf Stream or Kuroshio axis (§§6.4.3, 6.4.5).

A further modulator of marine CO_2 draw-down is the strength of deep convection in the North Atlantic. Most winters, over-turning transports large amounts of carbon to the intermediate and deep ocean. The location and strength of this convection is sensitive to the density of the surface waters, particularly through salinity variations (§6.4.5), and may have essentially stopped at times in the geological past (§6.2.2). The recent cooling of the northern North Atlantic should increase the density of the sub-polar surface waters and thus be re-inforcing this convection (although perhaps changing its location), and hence CO_2 draw-down. However,

Table 7.2. *Typical albedos of various surfaces*

Surface	Condition	Albedo
Snow	fresh	0.95
	old	>0.40
Ice	sea	0.30–0.45
	glacier	0.20–0.40
Water	small zenith angle	0.03–0.10
	large zenith angle	0.10–1.00
Soils	bare	0.05–0.40
Sand		0.20–0.45
Grass		0.16–0.26
Tundra		0.18–0.25
Forests	deciduous	0.15–0.20
	coniferous	0.05–0.15

Fig. 7.13. The latitude of the boundary of the polar ice cap as a function of solar constant, from an energy budget climate model. The model assumes an ice-free albedo of 0.3 and an ice-albedo of 0.62. [Fig. 9.5 of Hartmann (1994), *Global physical climatology.* Reproduced with kind permission of Academic Press.]

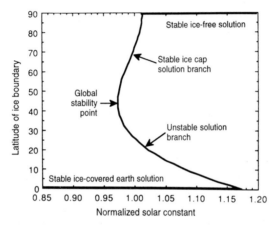

possible cycles in convective strength (§6.4.5) and limited observational data leave the question open. The various mechanisms affecting CO_2 draw-down are summarised in Fig. 7.12.

Deep water is also formed from the expulsion of salt during the freezing of sea-ice (§1.3.2). The abundance of sea-ice and land glaciers will be affected by the enhanced greenhouse effect. A warming troposphere will melt land-ice in marginal climates and make the formation of sea-ice more difficult. The result would be a decrease in the global area of ice. This has a strong positive feedback on climate because ice and snow have the highest albedos of almost any natural surface (Table 7.2). Decreasing ice cover greatly increases the amount of solar radiation that can be absorbed by the land or ocean surface. The ice-albedo feedback has a global impact but the regional amplification of warming is even stronger, as will be seen in §7.3.2. The sensitivity of global climate to this feedback can be seen in a simple energy budget model allowing variation in the solar constant (note that the model ignores the important effects of atmospheric and oceanic dynamics in transporting heat; Fig. 7.13)[4].

[4] Note that more realistic energy budget models push the stability point of Fig. 7.13 equatorwards, making it unlikely that the Earth has ever achieved sufficient ice cover to produce a run-away positive feedback forcing global glaciation.

Enhancement of the greenhouse effect acts to effectively increase the solar constant, radiatively by about 1% by 1990, ignoring other positive or negative feedbacks. Ice cover variations will in turn determine the atmospheric (and oceanic) circulation. However, apart from some evidence of net glacial melt contributing slightly to current sea level rise there is little confirmed sign of ice cover responding to the enhanced greenhouse effect (§6.4.2).

Other, lesser, feedbacks merit notice. Although a weaker polar–equator temperature gradient, and thus driving mechanism for the atmospheric circulation, would be predicted from ice-albedo feedback a greater quantity of latent heating provides more energy for a more vigorous and stormy circulation. Suggestive evidence for local marine wind increase has been presented earlier (§§6.4.3, 6.4.5). Analogously, greater latent heating in the tropics could lead to more frequent and severe hurricanes; some uncertain support is available for this (§6.4.2). Greater storminess, of course, is likely to increase thick cloud cover and may result in a negative feedback.

The burgeoning of ultra-violet light levels resulting from decaying stratospheric ozone may do more than add to the tropospheric energy bank. Ultra-violet light causes cell mutation and, potentially, death to many organisms. Plankton may be reduced significantly in number by increasing levels of this radiation, although ultra-violet light is absorbed rapidly in the top few metres of the sea (§2.1.1). This may be particularly severe during the spring blooms of productive sub-polar seas beneath the growing spring polar ozone holes. The net feedback would be positive because of a slowing of the oceanic carbon sink. Currently the only evidence for increased levels of ultra-violet light affecting marine organisms is strengthening of cell walls, and paradoxical utilisation of the extra solar energy for photosynthesis. This feedback remains speculative.

Global mean sea level has risen by about 11 cm this century. About a third of this rise is due to the expansion of sea water accompanying the net global warming over that time (Fig. 6.26). The rest is probably due to melting of small glaciers and the Greenland ice sheet, although probably not Antarctica. In fact some estimates of ice cap change suggest that Antarctica has accumulated volume this century. The recent cooling over the North Atlantic, and increased precipitation through stronger westerlies (§6.4.5), may now be leading to net accumulation over Greenland as well. Recent discussion has suggested that sea level would have risen faster, by another 2–4 cm, without considerable containment of water in recently built dams. The climatic effect of rising sea level is small. Any resulting increase in ocean surface area will have decreased the global albedo but the effect will have been negligible.

Changes to the terrestrial albedo by land use are probably small, but alteration to terrestrial ecosystems by enhancement of atmospheric CO_2 could result in climatic feedbacks. The raised CO_2 should encourage photosynthesis and enlarge the terrestrial carbon repository, although higher temperatures act against this by encouraging plant respiration at the expense of photosynthesis. The net effect is not known, although it seems that some CO_2 must be presently taken up by terrestrial organisms (§7.2.1).

7.3 The climate of the future

Natural climatic change occurs over many timescales. Astronomical forcing suggests that the planet is coming to the end of the current interglacial, with the prospect of mild global cooling over the next 10 000–20 000 years (§7.1.2). Significant fluctuations in climate over the last thousand years are poorly understood; it is doubtful that they are linked to humanity so such perturbations cannot be discounted over the next millenium (§§7.1.1, 7.1.5). The warming of the last century, however, represents both a problem and a challenge. Anthropogenic changes to the environment are indisputable (§§7.2.1–7.2.3). Their potential climatic impact is significant but the complex feedbacks within the climate system, not all of which may yet be known let alone understood, make it impossible, as of the mid-1990s, to categorically ascribe this warming to humanity's influence. There are too many positive and negative feedbacks whose true sizes, and even signs in some cases, are unknown (§7.2.4).

Rising levels of greenhouse gases are, however, likely to continue. All estimates of the CO_2 annual emission rate assuming realistic global population growth over the next century predict emissions to at least triple by 2100 (Fig. 7.14). These estimates all assume at least partial international compliance with projected emission controls. The Montreal Protocol, and its subsequent revisions, for the elimination of CFC emissions should eventually reverse the stratospheric loss of ozone, although not until well into the twenty-first century. The replacement chemicals are, however, also greenhouse gases, if less harmful to the ozone layer, and will contribute to an enhanced greenhouse effect. Effective doubling of the pre-industrial levels of CO_2, allowing for the contribution of other non-water vapour gases, is likely by 2025. This would result in a further decrease of 1.5 Wm^{-2} in the infra-red flux upward through the tropopause, supplementing the existing 2.5 Wm^{-2}.

Growing radiative forcing through escalation of the existing enhanced greenhouse effect therefore provides a mechanism for anthropogenic climatic change over the next century. The planet's ecosystem has coped with such change before, although probably not so rapidly. However, humanity's domination of the global environment creates an unprecedented situation. The sheer numerical size of our species, our political system of nation states with populations historically attached to their borders and identities, and the absence of clear international decision-making suggests the probability of future conflict if climatic boundaries shift appreciably and quickly.

Prediction of the timing and magnitude of climatic change over the next century is critical for policies of adaptation or emission control to be prepared, if necessary. The main tools for climate forecasting are numerical models of the general circulation of the atmosphere and ocean (§7.3.1). Many groups around the world have attempted to predict the equilibrium climate of an atmosphere with double the present levels of CO_2 (§7.3.2). Perhaps more important is estimating the speed with which the climate system will adjust to changing CO_2 levels: the so-called *transient* response

Fig. 7.14. Possible
scenarios of annual CO_2
emissions into the
atmosphere from tropical
deforestation, and energy
and cement production.
[Fig. A3.1 of Houghton *et
al.* (1992). Reproduced with
permission of Cambridge
University Press.]

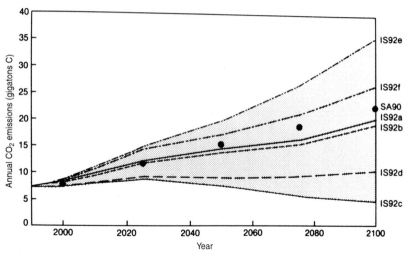

Fig. 7.15. Allocation of
model variables in space
within a typical ocean
GCM. Note that
temperature (T), salinity (S)
and streamfunction (ψ) are
staggered relative to the
velocity vector (V). [Fig. 1
of Han (1988). Reproduced
with kind permission of
Kluwer Academic
Publishers.]

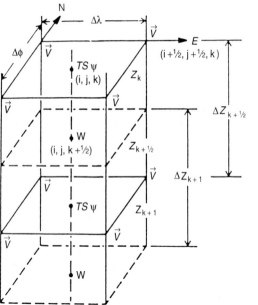

(§7.3.3). Numerical models, however, are far from perfect. Accurate inclusion of the many feedbacks discussed in §7.2.4, even those whose physical, chemical and biological mechanisms are well understood, is problematic because of the contrasting spatial scales of the model and processes. Early detection of climatic change which can unequivocally be attributed to the enhanced greenhouse effect and related to model predictions is therefore the key to validating current forecasts (§7.3.4).

7.3.1   Numerical models of the atmosphere and ocean

Circulation in the ocean and atmosphere can be described by well established sets of mathematical equations valid at every point in each fluid.

They describe the evolution of the density field and velocity in each direction. However, except in very special situations not encountered in the real world, exact solutions to these equations cannot be found. Techniques for numerically approximating the equations were developed as early as the First World War by L. F. Richardson, who attempted the first, unsuccessful, numerical weather prediction while driving a military ambulance in the battlefields of France. The advent of electronic computers in the 1940s and 1950s allowed numerical solution of these equations to become technically feasible.

The equations of motion cannot be numerically solved at every point in space – an infinite set – so an approximation to these equations is made by solving for the essential variables of velocity, temperature, salinity (in the ocean) and surface pressure and humidity (in the atmosphere) at discrete points of a grid in space[5] at discrete moments in time. Approximation to the terms in the equations are then made, linking the values at neighbouring grid points. A typical grid network for an ocean *general circulation model* (GCM) is shown in Fig. 7.15. Typical climate models currently have horizontal resolutions – the separation between grid points – of 3–5° and vertical resolutions of several hundred metres in the ocean and a kilometre or two in the atmosphere, while timesteps are tens of minutes (atmospheric GCMs) to hours (oceanic GCMs). These parameters are dictated by the memory and disk capability of current computers, and are continually reducing. In atmospheric GCMs it is often numerically advantageous to rewrite the vertical variation of the equations in terms not of altitude but a coordinate that smoothly follows the surface.

Unfortunately, the climate system involves many more processes than just atmospheric and oceanic motion (see Fig. 1.1). Radiation absorption and emission, cloud formation and precipitation, chemical reactions in the atmosphere, biological processes in the ocean, sea-ice formation, deep convection, snow melt and river run-off, the transfer of particles and gases between ocean and atmosphere, the storage and release of heat and moisture from terrestrial surfaces and vegetation: all of these processes occur on fundamentally smaller spatial scales than are likely to be resolvable within models for the foreseeable future. Often, as we have seen, the basic mechanisms driving such processes are not well understood. These physical, chemical and biological aspects of the climate system must therefore be simplified in order to be represented on GCM grids. This is known as *parameterisation*. Parameterisation schemes differ between groups of climate modellers and pose a crucial uncertainty in current predictions. Almost no current climate models include marine biological processes.

Most climate model experiments use either an atmospheric GCM or an ocean GCM, but rarely have the two been coupled together fully, at least until the mid 1990s. This is because of the doubling of computer resources required to do so, and the tendency for such coupled models to drift away from the stable climates that the individual components attain independently. Making the transfer of moisture, heat and momentum truly

---

[5] Other techniques can be used, such as a *spectral* formulation, but for our purposes these are equivalent to the standard approximation.

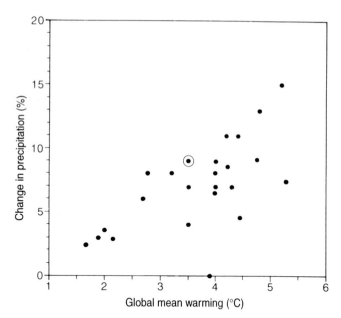

Fig. 7.16. The equilibrium response of a number of different climate models to a doubling of atmospheric concentrations of CO_2. This is shown in terms of the rise in global mean surface air temperature and global precipitation. Experiments up to 1992 are included. The circled experiment of the UK Meteorological Office is used for more detailed study in later figures. Data from Houghton *et al.* (1990, 1992).

interactive currently eludes climate modellers, although it is possible to reduce the climate drift problem by imposing a *flux correction* – an essentially artificial removal or addition of energy – to keep the coupled model under control. Having to resort to such a technique makes interpretation of coupled climatic change experiments problematic, although the consistency of transient response simulations between different models with and without flux correction is encouraging.

7.3.2 *Climate with doubled CO_2*

The benchmark for studies of the climatic impact of an enhanced greenhouse effect is the climate system's response to a doubling of the radiative impact of CO_2. The *equilibrium* response assumes that the climate system has had time to settle down to an equilibrium climate after the greenhouse gases have been added to the atmosphere. In reality, a state of equilibrium with the probable continual rise of emissions (Fig. 7.14) is unlikely to be attained for many decades, if not centuries. However, testing model responses to this idealised doubling helps to establish bounds on the sensitivity of the climate.

A single number which characterises this climate sensitivity is the equilibrium global surface air temperature change from doubling CO_2. There is scatter in the two dozen or so such atmospheric model experiments so far carried out (Fig. 7.16), none with sulphate aerosol feedback or biological production in their (rare) oceans, but all produce a warming, with more recent models, containing better parameterisations, giving a sensitivity range of 1.5°–4.5°C. An accompanying increase in global precipitation, because of the greater concentrations of water vapour, of 3–10% is predicted by the models.

Fig. 7.17. Change in
Northern Hemisphere
winter surface air
temperature, averaged over
10 years, in the high
resolution UK
Meteorological Office's
equilibrium response to
doubled CO_2, relative to
1951–80. Contours are
shown every 2°C, light
stippling shows rises in
temperature of more than
4°C, dashed shading of
more than 8°C. [From Fig.
5.4 of Houghton *et al.*
(1990). Reproduced with
permission of Cambridge
University Press.]

DJF 2 × CO2 - 1 × CO2 Surface air temperature: UKHI

Fig. 7.18. Change in
Northern Hemisphere
summer surface air
temperature, averaged over
10 years, in the high
resolution UK
Meteorological Office's
equilibrium response to
doubled CO_2, relative to
1951–80. Contours are
shown every 2°C, light
stippling shows rises in
temperature of more than
4°C, dashed shading of
more than 8°C. [From Fig.
5.4 of Houghton *et al.*
(1990). Reproduced with
permission of Cambridge
University Press.]

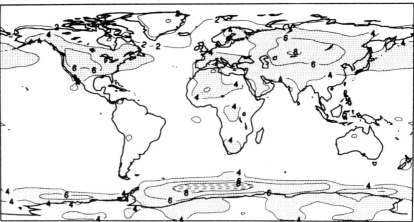

JJA 2 × CO2 - 1 × CO2 Surface air temperature: UKHI

Concurrence of model estimates of the global response of the climate to a doubling of CO_2 does not mean that predictions for regional changes are also similar. The resolution and differing parameterisation techniques of climate models mean that little credence can currently be given to the magnitude of regional predictions. The different models do, however, have certain large-scale characteristics in common. Predictions using the high resolution model (2.5° × 3.75°) of the U.K. Meteorological Office will be discussed, as a representative example from the centre of Fig. 7.16.

The major temperature response of all models is in high latitudes during the winter. This is particularly so in the northern hemisphere (Fig. 7.17), both over the Arctic where the sea-ice retreats significantly, and over continental interiors receiving less (high albedo) snowfall. The Southern Hemisphere winter (Fig. 7.18) experiences a smaller temperature rise, due to a relatively small decline in sea-ice extent. The interiors of the Northern Hemisphere land masses also warm appreciably in summer. The tropics show significantly smaller temperature increases throughout the year.

Fig. 7.19. Change in
Northern Hemisphere
winter precipitation,
averaged over 10 years, in
the high resolution UK
Meteorological Office's
equilibrium response to
doubled CO_2, relative to
1951–80. Contours are
shown at 0, ± 1, ± 2, ± 5
mm/day, light stippling
shows areas of decrease.
[From Fig. 5.6 of
Houghton *et al.* (1990).
Reproduced with
permission of Cambridge
University Press.]

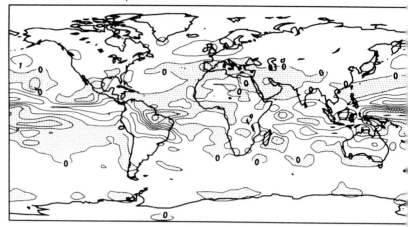

DJF 2 × CO2 - 1 × CO2 Precipitation: UKHI

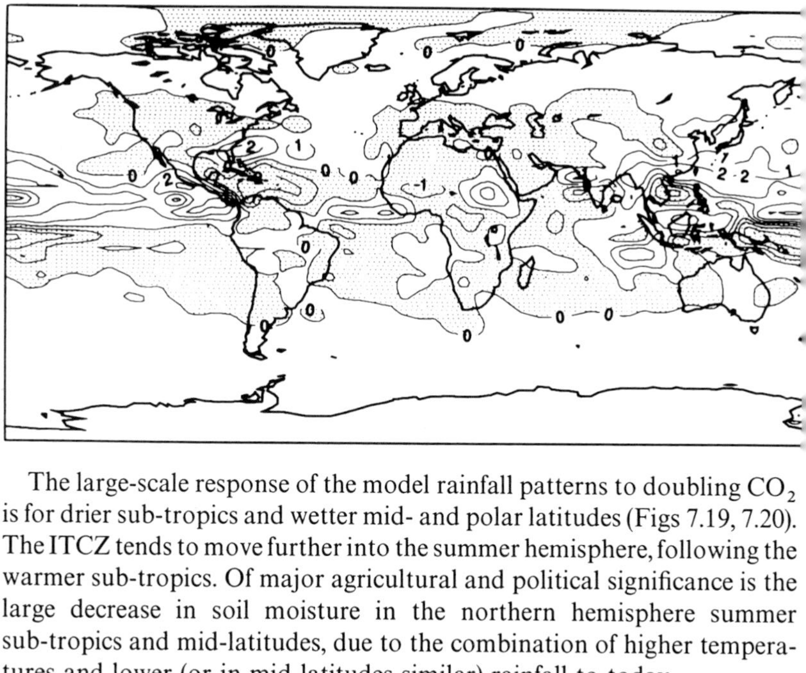

Fig. 7.20. Change in
Northern Hemisphere
summer precipitation,
averaged over 10 years, in
the high resolution UK
Meteorological Office's
equilibrium response to
doubled CO_2, relative to
1951–80. Contours are
shown at 0, ± 1, ± 2, ± 5
mm/day, light stippling
shows areas of decrease.
[From Fig. 5.6 of
Houghton *et al.* (1990).
Reproduced with
permission of Cambridge
University Press.]

JJA 2 × CO2 - 1 × CO2 Precipitation: UKHI

The large-scale response of the model rainfall patterns to doubling CO_2 is for drier sub-tropics and wetter mid- and polar latitudes (Figs 7.19, 7.20). The ITCZ tends to move further into the summer hemisphere, following the warmer sub-tropics. Of major agricultural and political significance is the large decrease in soil moisture in the northern hemisphere summer sub-tropics and mid-latitudes, due to the combination of higher temperatures and lower (or in mid-latitudes similar) rainfall to today.

7.3.3 Modelling the transient response to CO_2 increase

In reality the climate system will be responding over the next century to constantly changing concentrations of greenhouse gases. Of more significance for forecasting climatic change therefore is the evolving response of the system to this forcing. The oceans are of particular importance during this type of climatic forcing because of their large capacity to absorb heat and act as a retarder to change. Climate models coupling oceanic and atmospheric GCMs are vital for predicting the climate's transient response.

Fig. 7.21. Variation of anomalous global mean surface air temperature with time in several coupled ocean–atmosphere model experiments. The emission rates differ from model to model, and the effect of climate drift has been removed. The 1990 IPCC 'best estimate' of likely global temperature change in the future is shown as the thick line. [Fig. B1 of Houghton *et al.* (1992). Reproduced with permission of Cambridge University Press.]

Fig. 7.22. Average surface air temperature anomaly during years 65–75 of the UK Meteorological Office simulation of Fig. 7.21. Contours are in °C, with areas of temperature decrease stippled. [Fig. B4d of Houghton *et al.* (1992). Reproduced with permission of Cambridge University Press.]

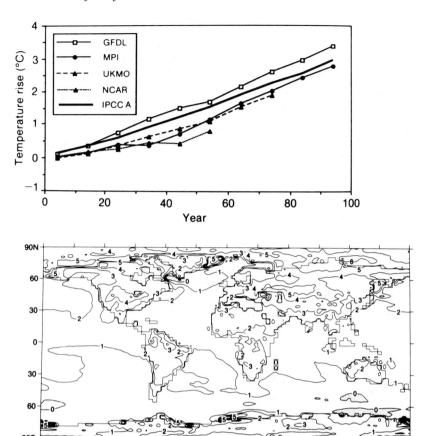

Several such models have been run. They employ different numerical methods and parameterisations, are of different resolution, take different starting values for the atmospheric CO_2 concentration and are thus of different climate sensitivity. This sensitivity, the equilibrium double CO_2 response, varies from 2.6° to 4.5°C. The models, adding CO_2 to the atmosphere at a rate of roughly 1% per year, show similar, linear, responses (Fig. 7.21). Each shows an increase in global mean surface temperature at the time of a doubling of CO_2 of 50–60% of the equilibrium value.

Climatic change is therefore likely to be less dramatic than the equilibrium experiments (§7.3.2) suggest. The speed with which different parts of the globe approach their equilibrium response differs considerably. The U.K. Meteorological Office mean surface temperature, relative to a present-day equilibrium, over the 10 years about the time of reaching double the initial CO_2 is shown in Fig. 7.22. While the change in polar regions is greater than in the tropics, the latter have almost attained their equilibrium response of Figs 7.17 and 7.18. The North Atlantic, and parts of the Southern Ocean, are particularly slow to respond to the higher CO_2, in all models. This is probably due to deep water formation transferring heat rapidly into the interior of the ocean.

Fig. 7.23. Present-day climate simulation by climate models. The zonally-averaged Northern Hemisphere summer sea level pressure, in mb, is shown in (a) and the zonally-averaged Northern Hemisphere winter precipitation, in mm/day, is shown in (b). Different models have different symbols; the solid line is the observed climatology. [Fig. B27 of Houghton *et al.* (1992). Reproduced with permission of Cambridge University Press.]

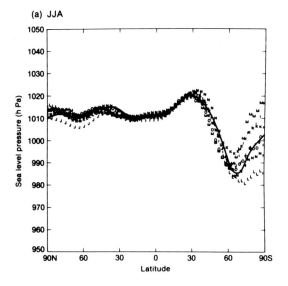

(a) JJA

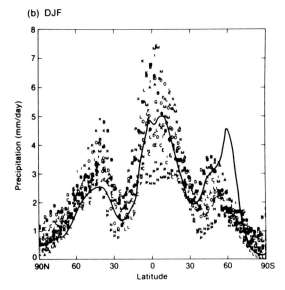

(b) DJF

7.3.4   Detection of climatic change

Both the equilibrium and transient model experiments of potential future climates suggest changes which can be sought in the observational record. Validation of the models' ability to reproduce both the present climate and climatic evolution is essential for belief in their predictions. Models are currently moderately successful in reproducing the present climate, but certainly not without blemish. Precipitation and the circulation in the presence of topography are not well reproduced (Fig. 7.23). Recent climatic change was shown in §§6.4 and 7.1.4 to be within the range of natural variability; coupled models confirm that the magnitude of natural variability in global mean surface temperature can be several tenths of a degree.

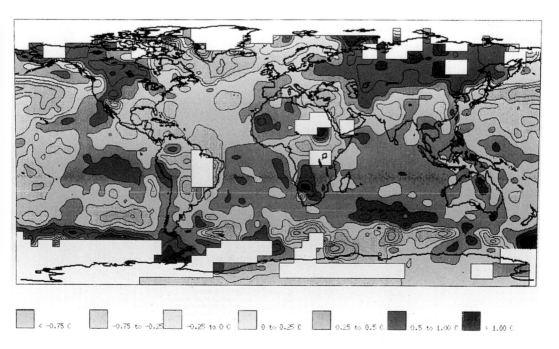

< -0.75 C	-0.75 to -0.25	-0.25 to 0 C	0 to 0.25 C	0.25 to 0.5 C	0.5 to 1.00 C	> 1.00 C

Fig. 7.24. Anomaly in annual mean surface air temperature during the decade 1981–90, relative to 1951–80. Squared regions show regions of no data. [Plate A of Houghton *et al.* (1990). Reproduced with permission of Cambridge University Press.]

Transient experiments, however, suggest that CO_2-driven climatic change of a magnitude exceeding probable natural variability should be imminent. Comparison of recent global climate with model predictions is therefore entering a crucial phase over the next 20 years. In Fig. 7.24 the annual surface temperature anomalies of the last complete decade, 1981–90, are compared to the mean of the 30 year period 1951–80. Comparison of this diagram with the transient response upon reaching double CO_2 (Fig. 7.22) is suggestive. Recent climatic change has seen warming over Northern Hemisphere continental interiors and some polar regions, particularly in winter. The sub-tropics are also warmer than in the middle of the century, although oceanic regions where deep water formation or subduction occur have tended to cool.

There is intriguing similarity between the patterns of climatic change found in the transient models and recent observations. However, this correspondence is not yet of sufficient size or permanence to prove that significant climatic change due to enhancement of the greenhouse effect is underway. There are still many inadequacies in model formulations, leading to differences between models and the absence of potentially crucial feedbacks from all of the climate models. These missing factors, or processes of which only the rudiments of understanding or model implementation are just now under trial in the mid 1990s, are more than minor. They include such effects as the addition of anthropogenically derived sulphate aerosols, the role of the ocean in absorbing CO_2 and the interactive feedbacks involving marine organisms, and atmospheric chemistry describing particulate formation in the troposphere and ozone chemistry in the stratosphere. These have been seen in §7.2.4 to contribute potentially large positive or negative feedbacks to the climate system which may change future model predictions radically.

In addition, the capability of the climate system to change rapidly due to quite minor perturbations within the system has been seen on numerous occasions in the last two chapters. These, perhaps unpredictable, possibilities lie hidden ready to derail the most accurate forecasts. The details of the climate, like the weather, are fundamentally unpredictable because of the chaotic character underpinning the climate system.

The study of climate approaches a crossroads. The impact of possible change has led to rapid advance in understanding the climate system. Simultaneously, the necessity arose to predict this change, while the fundamental improvement of our understanding proceeds. Modern climatologists may be on the verge of providing advice crucial for world peace and sustainable development in the twenty-first century. Nonetheless, the possibility remains that years of public warning will prove unwarranted. The probability of this is perhaps slight, but the complexities of only the ocean's role in climate, hopefully unravelled in this book, show what uncertainty remains in understanding the oceans and climate.

Further reading

A complete reference list is available at the end of the book but the following is a selection of the best books or articles to follow up particular topics within this chapter. Full details of each reference are to be found in the Bibliography.

Hartmann (1994): A well written text on modern climatology that examines many of the physical feedbacks within the climate system.

Houghton *et al.* (1990): Careful discussion of the evidence for current and future climatic change. The definitive presentation of the current state of climatic research.

Houghton *et al.* (1992): A supplement to the 1990 report, with useful new evidence for climatic change.

Houghton *et al.* (1995): Yet another volume from the Intergovernmental Panel for Climatic Change. This one examines the various aspects of radiative forcing in detail. It also evaluates future emission scenarios for trace gases.

Kemp (1994): A good, basic discussion of a number of global climatological issues, with the emphasis on the climatological and policy implications of each.

Appendix A Useful constants and the electromagnetic spectrum

Solar constant	$S = 1380 \text{ Wm}^{-2}$
Stefan–Boltzman constant	$\sigma = 5.67 \times 10^{-8} \text{ Wm}^{-2}\text{K}^{-4}$
Planck's constant	$\hbar = 6.63 \times 10^{-34} \text{ kgm}^2\text{s}^{-1}$
Speed of light	$c = 3.00 \times 10^8 \text{ ms}^{-1}$
Average distance of Earth from Sun	$D = 1.50 \times 10^8 \text{ km}$
Gravity at sea level	$g = 9.81 \text{ ms}^{-2}$
Mean radius of the Earth	$R = 6370 \text{ km}$
Earth's angular speed of rotation	$\Omega = 7.292 \times 10^{-5} \text{ rad s}^{-1}$
Mass of the Earth	$M = 5.988 \times 10^{24} \text{ kg}$
Albedo of the Earth	$a = 0.30$
Equivalent temperature of Earth	$T_E = 255 \text{ K}$
Universal gas constant	$R^* = 8.314 \times 10^3 \text{ JK}^{-1}\text{mol}^{-1}$
Gas constant for dry air	$R = 287 \text{ JK}^{-1}\text{kg}^{-1}$
Avogadro constant	$N_0 = 6.02 \times 10^{23} \text{ mol}^{-1}$
Specific heat of dry air at constant pressure	$c_p = 1004 \text{ JK}^{-1}\text{kg}^{-1}$
Specific heat of water at constant pressure	$c_{pw} = 4217.7 \text{ JK}^{-1}\text{kg}^{-1}$
Latent heat of evaporation at 0°C	$L_v = 2.5 \times 10^6 \text{ Jkg}^{-1}$
Latent heat of melting at 0°C	$L_s = 3.3 \times 10^5 \text{ Jkg}^{-1}$
Standard sea-level atmospheric pressure	$p_0 = 101325 \text{ Pa}$
Standard sea-level atmospheric temperature	$T_0 = 288.15 \text{ K}$
Standard sea-level atmospheric density	$\rho_a = 1.225 \text{ kgm}^{-3}$
Standard sea-level ocean density	$\rho_w = 1026 \text{ kgm}^{-3}$
Refractive index of air	$n_a = 1.00$
Refractive index of water	$n_w = 1.33$

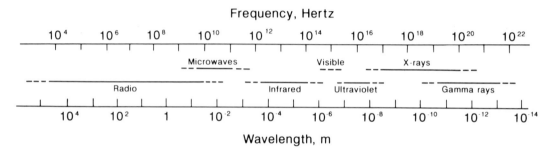

Fig. A.1. Electromagnetic
spectrum. Frequency
shown above, wavelength
below, with approximate
extents of different types of
bands.

Appendix B *Periodic table and electron orbital configuration*

In order to understand qualitatively some of the radiative and chemical processes discussed in the text a general appreciation of the structure of the electron cloud around atoms is necessary. Table B.1 gives the Periodic Table of the naturally occurring elements and Table B.2 shows the configuration of the electrons in the orbitals surrounding the appropriate atomic nucleus. The reader will probably be aware of the similarity of chemical properties exhibited by the elements in vertical columns of the Periodic Table. Table B.2 suggests that this has much to do with the electronic structure.

In Chapter 1 we encountered the concept of energy within atoms or molecules being quantised, or only allowed a discrete number of values. The theory of quantum mechanics extends the ideas of classical mechanics to such systems. In 1926 Schrödinger developed an equation to describe the position and energy of the electron around the hydrogen atom, using the concept that the total energy is the sum of kinetic and potential energies. The solution to this equation allows only discrete energies, and electron orbital shapes, to exist. The orbitals are not the exact tracks which the electron may follow but give the region of space around the nucleus where the electron is most likely to be found when it possesses a particular energy.

The solution to the Schrödinger equation for the hydrogen atom has three parameters, or quantum numbers, which determine the electron configuration. One (the shell noted in Table B.2) describes the effective volume of an electron orbital; a second (the s, p, d and f parameters) the basic shape of the region where the electron is most likely to be found in the orbital; and the third determines the orientation of the particular orbital shape. In addition, there is a fourth quantum number for the spin of the electron, arising from its charged nature, which can only have two states. The particular configuration for an atom arises because of the Pauli Exclusion Principle, which states that no two electrons around a nucleus may possess an identical set of quantum numbers. This principle is a result of the repulsive potential of two such electrons due to their charge.

Hydrogen is a very simple atom as it possess only one proton in the nucleus and one orbiting electron. Other elements possess more complicated structures, but exhibit the same basic electron orbitals as hydrogen. Molecules possess electron orbitals which are essentially geometric combinations of those of their constituent atoms. The energy levels of the basic shells for neutral many-electron atoms are shown in Figure B.1.

Comparison of Table B.2 with Fig. B.1 shows that the filling of electron orbitals follows the principal of minimum energy. In the main text of the book we encounter situations, such as in the Chapman reactions of ozone in the stratosphere (equation (1.4)), where atoms, or molecules, are excited. In this situation one of their electrons is temporarily displaced into a higher energy state than is found in the basic configuration.

Table B.1. *Naturally occurring elements*

Key: from top to bottom: Atomic number, name, chemical symbol, atomic weight of principal isotope. Atomic weight displayed in [] indicates the principal isotope is radioactive

1 Hydrogen H 1.008								
3 Lithium Li 6.94	4 Berylium Be 9.01							
11 Sodium Na 22.99	12 Magnesium Mg 24.31							
19 Potassium K 39.01	20 Calcium Ca 40.08	21 Scandium Sc 44.96	22 Titanium Ti 47.90	23 Vanadium V 50.94	24 Chromium Cr 52.00	25 Manganese Mn 54.94	26 Iron Fe 55.85	27 Cobalt Co 58.93
37 Rubidium Rb 85.47	38 Strontium Sr 87.62	39 Yttrium Y 88.91	40 Zirconium Zr 91.22	41 Niobium Nb 92.91	42 Molybdenum Mo 95.94	43 Technetium Tc 98.91	44 Ruthenium Ru 101.07	45 Rhodiur Rh 102.91
55 Caesium Cs 132.91	56 Barium Ba 137.34	57-71 Lanthan- ides	72 Hafnium Hf 178.49 73	73 Tantalum Ta 180.95	74 Tungsten W 183.85	75 Rhenium Re 186.2	76 Osmium Os 190.2	77 Iridium Ir 192.22
87 Francium Fr [223]	88 Radium Ra [226]	89-103 Actinides						

LANTHANIDES		57 Lanthanum La 138.91	58 Cerium Ce 140.12	59 Praseodymium Pr 140.91	60 Neodymium Nd 144.24	61 Promethium Pm [145]	62 Samarium Sm 150.4	63 Europiu Eu 151.96
ACTINIDES		89 Actinium Ac [227]	90 Thorium Th 232.04	91 Protoactinium Pa 231.04	92 Uranium U 238.03			

Fig. B.1. Relative energies of the orbitals in neutral many-electron atoms. In building the periodic table (Table B.1) electrons are put into the lowest energy states first, so, for instance, the 4s orbital is filled before the 3d (see Table B.2). The variable *n* indicates the shell number.

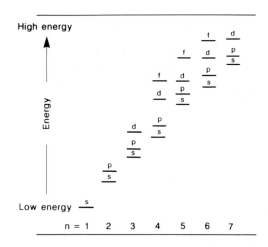

			5	6	7	8	9	2 Helium He 4.00 10
			Boron	Carbon	Nitrogen	Oxygen	Fluorine	Neon
			B	C	N	O	F	Ne
			10.81	12.01	14.01	16.00	19.00	20.18
			13	14	15	16	17	18
			Aluminium	Silicon	Phosphorus	Sulphur	Chlorine	Argon
			Al	Si	P	S	Cl	Ar
			26.98	28.09	30.97	32.06	35.45	39.95
28	29	30	31	32	33	34	35	36
Nickel	Copper	Zinc	Gallium	Germanium	Arsenic	Selenium	Bromine	Krypton
Ni	Cu	Zn	Ga	Ge	As	Se	Br	Kr
58.71	63.55	65.37	69.72	72.59	74.92	78.96	79.90	83.80
46	47	48	49	50	51	52	53	54
Palladium	Silver	Cadmium	Indium	Tin	Antimony	Tellerium	Iodine	Xenon
Pd	Ag	Cd	In	Sn	Sb	Te	I	Xe
106.4	107.87	112.40	114.82	118.69	121.75	127.60	126.90	131.30
78	79	80	81	82	83	84	85	86
Platinum	Gold	Mercury	Thallium	Lead	Bismuth	Polonium	Astatine	Radon
Pt	Au	Hg	Tl	Pb	Bi	Po	At	Rn
195.09	196.97	200.59	204.37	207.2	208.98	[209]	[210]	[222]

64	65	66	67	68	69	70	71
Gadolinium	Terbium	Dysprosium	Holmium	Erbium	Thulium	Ytterbium	Lutetium
Gd	Tb	Dy	Ho	Er	Tm	Yb	Lu
157.25	158.93	162.50	164.93	167.26	168.93	173.04	174.97

Table B.2. *Electron orbital configuration for natural elements.*
Within each shell there are s, p, d and f orbitals

Element	Orbital configuration (shell)						
	1	2	3	4	5	6	7
Hydrogen	1						
Helium	2						
Lithium	2	1					
Berylium	2	2					
Boron	2	2 1					
Carbon	2	2 2					
Nitrogen	2	2 3					
Oxygen	2	2 4					
Fluorine	2	2 5					
Neon	2	2 6					
Sodium	2	2 6	1				
Magnesium	2	2 6	2				
Aluminium	2	2 6	2 1				
Silicon	2	2 6	2 2				
Phosphorus	2	2 6	2 3				
Sulphur	2	2 6	2 4				
Chlorine	2	2 6	2 5				
Argon	2	2 6	2 6				
Potassium	2	2 6	2 6	1			
Calcium	2	2 6	2 6	2			
Scandium	2	2 6	2 6 1	2			
Titanium	2	2 6	2 6 2	2			
Vanadium	2	2 6	2 6 3	2			
Chromium	2	2 6	2 6 5	1			
Manganese	2	2 6	2 6 5	2			
Iron	2	2 6	2 6 6	2			
Cobalt	2	2 6	2 6 7	2			
Nickel	2	2 6	2 6 8	2			
Copper	2	2 6	2 6 10	1			
Zinc	2	2 6	2 6 10	2			
Gallium	2	2 6	2 6 10	2 1			
Germanium	2	2 6	2 6 10	2 2			
Arsenic	2	2 6	2 6 10	2 3			
Selenium	2	2 6	2 6 10	2 4			
Bromine	2	2 6	2 6 10	2 5			
Krypton	2	2 6	2 6 10	2 6			
Rubidium	2	2 6	2 6 10	2 6	1		
Strontium	2	2 6	2 6 10	2 6	2		
Yttrium	2	2 6	2 6 10	2 6 1	2		
Zirconium	2	2 6	2 6 10	2 6 2	2		
Niobium	2	2 6	2 6 10	2 6 4	1		
Molybdenum	2	2 6	2 6 10	2 6 5	1		
Technetium	2	2 6	2 6 10	2 6 6	1		
Ruthenium	2	2 6	2 6 10	2 6 7	1		
Rhodium	2	2 6	2 6 10	2 6 8	1		
Palladium	2	2 6	2 6 10	2 6 10			
Silver	2	2 6	2 6 10	2 6 10	1		
Cadmium	2	2 6	2 6 10	2 6 10	2		
Indium	2	2 6	2 6 10	2 6 10	2 1		
Tin	2	2 6	2 6 10	2 6 10	2 2		

Table B.2. *(cont.)*.

Within each shell there are s, p, d and f orbitals

Element	Orbital configuration (shell)						
	1	2	3	4	5	6	7
Antimony	2	2 6	2 6 10	2 6 10	2 3		
Tellurium	2	2 6	2 6 10	2 6 10	2 4		
Iodine	2	2 6	2 6 10	2 6 10	2 5		
Xenon	2	2 6	2 6 10	2 6 10	2 6		
Calcium	2	2 6	2 6 10	2 6 10	2 6	1	
Barium	2	2 6	2 6 10	2 6 10	2 6	2	
Lanthanum	2	2 6	2 6 10	2 6 10	2 6 1	2	
Cerium	2	2 6	2 6 10	2 6 10 2	2 6	2	
Praseodymium	2	2 6	2 6 10	2 6 10 3	2 6	2	
Neodymium	2	2 6	2 6 10	2 6 10 4	2 6	2	
Promethium	2	2 6	2 6 10	2 6 10 5	2 6	2	
Samarium	2	2 6	2 6 10	2 6 10 6	2 6	2	
Europium	2	2 6	2 6 10	2 6 10 7	2 6	2	
Gadolinium	2	2 6	2 6 10	2 6 10 7	2 6 1	2	
Terbium	2	2 6	2 6 10	2 6 10 9	2 6	2	
Dysprosium	2	2 6	2 6 10	2 6 10 10	2 6	2	
Holmium	2	2 6	2 6 10	2 6 10 11	2 6	2	
Erbium	2	2 6	2 6 10	2 6 10 12	2 6	2	
Thulium	2	2 6	2 6 10	2 6 10 13	2 6	2	
Ytterbium	2	2 6	2 6 10	2 6 10 14	2 6	2	
Lutetium	2	2 6	2 6 10	2 6 10 14	2 6 1	2	
Hafnium	2	2 6	2 6 10	2 6 10 14	2 6 2	2	
Tantalum	2	2 6	2 6 10	2 6 10 14	2 6 3	2	
Tungsten	2	2 6	2 6 10	2 6 10 14	2 6 4	2	
Rhenium	2	2 6	2 6 10	2 6 10 14	2 6 5	2	
Osmium	2	2 6	2 6 10	2 6 10 14	2 6 6	2	
Iridium	2	2 6	2 6 10	2 6 10 14	2 6 7	2	
Platinum	2	2 6	2 6 10	2 6 10 14	2 6 9	1	
Gold	2	2 6	2 6 10	2 6 10 14	2 6 10	1	
Mercury	2	2 6	2 6 10	2 6 10 14	2 6 10	2	
Thallium	2	2 6	2 6 10	2 6 10 14	2 6 10	2 1	
Lead	2	2 6	2 6 10	2 6 10 14	2 6 10	2 2	
Bismuth	2	2 6	2 6 10	2 6 10 14	2 6 10	2 3	
Polonium	2	2 6	2 6 10	2 6 10 14	2 6 10	2 4	
Astatine	2	2 6	2 6 10	2 6 10 14	2 6 10	2 5	
Radon	2	2 6	2 6 10	2 6 10 14	2 6 10	2 6	
Francium	2	2 6	2 6 10	2 6 10 14	2 6 10	2 6	1
Radium	2	2 6	2 6 10	2 6 10 14	2 6 10	2 6	2
Actinium	2	2 6	2 6 10	2 6 10 14	2 6 10	2 6 1	2
Thorium	2	2 6	2 6 10	2 6 10 14	2 6 10	2 6 2	2
Protoactinium	2	2 6	2 6 10	2 6 10 14	2 6 10 2	2 6 1	2
Uranium	2	2 6	2 6 10	2 6 10 14	2 6 10 3	2 6 1	2

Appendix C *Stability, potential temperature and density*

The tendency for the atmosphere or ocean to mix vertically – its stability – is an important climatic parameter. The ease with which the ocean can spontaneously move vertically affects the rate and properties of deep water formation, upwelling and the rate of deepening or shallowing of the surface mixed layer. The stability of the atmosphere dictates the extent and depth of cloud formation, and thus precipitation. It also controls the ease of transferring gases and aerosols between the free atmosphere and the planetary boundary layer.

Adiabatic motion conserves energy. This means that the pressure, volume and temperature of a parcel of fluid moving vertically will vary merely because of the changing pressure of the surrounding environment on the parcel's envelope. In the atmosphere there is a simple equation of state relating these three variables:

$$P = \rho R T \tag{C.1}$$

where P is the pressure, R is the gas constant for dry air, ρ is the density and T is the temperature. A change in pressure causes an inverse change in volume of the parcel, and hence ρ (see equation (2.7)); temperature must also change to satisfy (C.1). It can be shown that for purely adiabatic vertical motion of air the temperature must change by 9.8°C per kilometre (the dry adiabatic lapse rate), as long as no condensation occurs. Potential temperature, θ, is defined as that temperature which a parcel of air would have if it was adiabatically moved to a pressure of 1000 mb. It can be expressed as:

$$\theta = T\left(\frac{1000}{P}\right)^{(R/c_{\mathrm{p}})} \tag{C.2}$$

where P is measured in millibars and c_{p} is the specific heat of dry air (see §1.3.1).

In the ocean, adiabatic vertical motion is less easily described because the effect of salinity on density leads to a complicated equation of state for sea water. The potential temperature of a parcel of water in the ocean, relative to the ocean surface, is only slightly different from the real temperature. The potential density relative to the surface (σ_θ), however, differs significantly from the actual density because of the oceanic equation of state's non-linear coupling of temperature, salinity and pressure.

Vertical stability in either fluid is dictated by the behaviour of a parcel of

248

fluid displaced adiabatically, by some mechanism, to a different height. Assume that the parcel has been displaced upwards and that temperature is being measured if the parcel is in the atmosphere but density if it is in the ocean. If the displaced fluid has adiabatically altered so as to be colder or denser than its surroundings it will sink towards its original position. Such fluid is said to be stable. If, however, the displaced fluid has adiabatically become warmer or lighter than its new surroundings then it will tend to rise further. Such fluid is unstable: initial movement is magnified by the vertical variation in the background environment. The displaced fluid may, of course, have exactly the same temperature or density as the environment. In this case the fluid has neutral stability: it will remain where it has been sent.

Stability in the atmosphere can be readily assessed from a vertical temperature sounding. Where the fall of temperature with height is less than 9.8°C per kilometre (or the temperature is rising) the air is stable and adiabatic motion will be suppressed. Contrastingly, if the fall of temperature with height is greater than 9.8°C per kilometre the air is unstable; adiabatic motion is encouraged. A complication to this simple picture occurs in the presence of clouds. Where condensation occurs in the atmosphere latent heat is released (§2.2.1), warming the environment via purely internal means. Adiabatic motion through a cloud layer will therefore change a parcel's temperature by less than in cloudless air because of this internal heating. The saturated adiabatic lapse rate varies somewhat with height and cloud properties, but is roughly 6.5°C per kilometre. There is thus a range of environments, with lapse rates between $9.8°\text{Ckm}^{-1}$ and $6.5°\text{Ckm}^{-1}$, which have conditional stability, that is, the air is stable if dry, but unstable if cloudy.

In the ocean this complication does not arise. Merely evaluating the potential density profile shows whether a parcel is stable or unstable. The non-linear equation of state means, however, that adiabatic motion is easily approximated only over relatively short depth ranges. Thus, for example, stability in the upper ocean is assessed by the potential density relative to the surface, but below 3000 m depth the potential density is better expressed relative to a deeper level, say 4000 decibars (approximately 4000 m).

Appendix D *Rossby waves in the atmosphere and ocean*

In the upper troposphere the length scales of the atmospheric circulation tend to be much greater than near the surface, as the influence of surface-related forcing is much reduced. The contrast between the warmth, and thickness, of the tropical troposphere and the cold, and shallowness, of the polar troposphere drives a strong westerly wind at mid-latitudes in this region of the troposphere. We saw in Chapter 2 that potential vorticity is conserved in rotating fluid flow; the effect of potential vorticity conservation on strong, large-scale winds is to produce large-amplitude waves as a result of any deflection from purely zonal flow. These waves are known as Rossby waves, after the first person who explored their significance.

In a quiescent atmosphere Rossby waves travel westward, with a wave speed, C, dependent on their wavelength, L:

$$C = -\frac{\beta L^2}{4\pi^2} \tag{D.1}$$

where $\beta = df/dy$ is the rate of change of the Coriolis parameter, f, with latitude. However, the upper westerlies are so strong that, relative to the Earth, the Rossby waves travel eastwards with the flow. Thus a series of snapshots of the upper atmospheric pressure field will show large-amplitude waves moving eastward, but at a speed somewhat slower than the actual peak wind speed. Near-surface atmospheric disturbances like frontal systems and depressions are embedded within this wave system and move with it, tending to be in the troughs of the upper tropospheric Rossby waves.

Such waves also occur in the ocean, but do not play such a significant role in oceanic circulation as in the atmosphere. Oceanic flow is so slow that oceanic Rossby waves (which tend to lie on the thermocline) travel westwards, as theory suggests. Thus Rossby waves can be observed travelling across the mid-latitude Pacific from North America to Asia. These may be involved in the periodicity of the ENSO cycle, as discussed in Chapter 5.

Glossary

absorption spectra	the variation with wavelength of *electromagnetic radiation* absorption by a molecule
adiabatic	a process not exchanging heat with the surroundings
aerosols	any small particles in the atmosphere
albedo	the proportion of incident energy reflected by an object, both directly and indirectly
anaerobic	the absence of oxygen
angular momentum	a conservative property of rotating bodies. *See* eq (2.9)
anion	a positively charged ion
anthropogenic	of human origin
anticyclone	a circulation system with high pressure or density at its core, rotating clockwise in the Northern Hemisphere and anticlockwise in the Southern
austral	southern
bathymetry	the topography of the sea floor
Bergeron–Findeisen process	formation of ice crystals by preferential vapour absorption
biome	a region of broadly similar vegetation and habitats
biosphere	the region of the Earth, both on land and in the ocean, in which life flourishes
black body	an object that absorbs all incident radiation, and emits radiation according to the *Stefan–Boltzmann Law*
boreal	northern
canonical	characteristic or typical
cation	a negatively charged ion
centre of mass	the point of balance of a system
centrifugal force	an apparent force needed to balance the centripetal force causing rotation, if viewed from the rotating object
chimney	a small region of *over-turning* in the ocean
chlorofluorocarbons	carbon compounds containing chlorine and fluorine
chlorophyll	one of the main by-products of *photosynthesis*
cirrus	high, thin clouds made of ice crystals
cloud amount	the proportion of sky covered by cloud, of whatever type
coalescence	formation of raindrops by collision of droplets
colloid	a large agglomeration of molecules or particles
conduction	transfer of heat by molecular collision
condensation nuclei	small particles in the atmosphere that act as initiators of condensation
constructive interference	the addition of two wave fields that amplifies the individual effects of each

conveyor belt	the theoretical route by which water circulates around the entire global ocean
Coriolis force	the apparent force needed to take into account the effect of the Earth's rotation on motion relative to a fixed point on its surface
coulombic forces	forces due to attraction and repulsion of electric charge
cross-isobaric	across, or at right angles to, isobars
cumulus	fluffy, bulbous cloud, mostly composed of water droplets and usually indicating unstable air
cumulonimbus	a cumulus cloud of great vertical extent containing both ice and water phases
cycling	the concept of substances continually moving from one storage region or reservoir to another, without a net gain or loss of the substance (usually an element)
cyclogenesis	the formation of a *cyclone*
cyclone	a circulation system with low pressure or density at its core, rotating anticlockwise in the Northern Hemisphere and clockwise in the Southern
declination	the tilt of the Earth's axis of rotation relative to the orbital plane
demineralization	conversion of *detritus* to nutrients by bacteria
detritus	dead organisms and remains of shells, etc. from the *plankton*
diabatic	a process involving exchange of heat with the surroundings
differential	an extremely small change, underlying the theory of calculus
dissociation constant	the rate at which a chemical reaction proceeds
Ekman layer	region of the upper ocean upon which the wind acts directly
Ekman pumping	the vertical motion induced by large-scale *Ekman transports*
Ekman spiral	The vertical shear of the directly wind-forced flow in the *Ekman layer*
Ekman transport	the net movement of water in the *Ekman layer*
electromagnetic radiation	radiation which possesses both electric and magnetic fields, e.g. visible light, X-rays, microwaves, radio waves, infra-red. *See* Fig. A.1
electron configuration	the arrangement of electrons into energy states within an atom
ENSO	El Niño *Southern Oscillation*
enthalpy of solution	the energy released when a mole of substance is dissolved in sea water
environmental lapse rate	the rate of change of temperature with height in the background air, *not* in an isolated parcel
equatorial Kelvin wave	a travelling disturbance in the equatorial upper ocean, caused by changes in surface winds, travelling eastwards with its maximum amplitude on the equator
equatorial Rossby wave	a disturbance similar to an *equatorial Kelvin wave* but travelling west rather than east, and most pronounced away from the equator

equilibrium response	the final climatic change occurring from alteration of one part of the climate system. Compare with the *transient response*
euphotic zone	the region of the upper ocean where the light level is greater than 1% of that at the surface
exothermic	a chemical reaction which releases heat
extra-tropics	the region of the globe poleward of 22.5°, but often taken to mean mid-latitudes
fall speed	the rate at which an object falls through the atmosphere, determined by a balance between gravity and air resistance
feedback	process by which a change caused by alteration in another quantity then alters the original variable. A positive feedback accentuates the original change, a negative one decreases it
fluorescence	conversion of short wave to longer wavelength radiation via jumps in electron energy states within an atom or molecule
flux correction	an alteration to the surface characteristics of a climate model to prevent its climate drifting over time
freon	*see* chlorofluorocarbons
friction velocity	a measure of the turbulence of the fluid
Gaia hypothesis	that biological processes interact with the Earth's climate to produce a life-sustaining environment
general circulation model (GCM)	a numerical model of the atmosphere or ocean containing representation of as many climatic processes as possible
geostrophy	the force balance between the pressure gradient and Coriolis forces
greenhouse effect	an increase in *tropospheric* temperature brought about by the absorption of infra-red radiation by particular, low concentration, gases in the atmosphere
gyre	a rotational circulation, usually in the ocean
Hadley cell	the basic large-scale, thermally-driven circulation of the atmosphere
harmonics	multiples of a basic frequency
heat budget	the proportioning of energy to various sources and sinks
heterogeneous	a chemical or physical process occurring on a surface
homogeneous	a chemical or physical process occurring in the absence of a surface
hydrostatic balance	the equilibrium state in a fluid where no vertical motion occurs as a result of a balance between gravity and the vertical pressure gradient
hygroscopic	attracts water vapour
Indo-Australian Convergence Zone	the region of atmospheric convergence situated over northern Australia or Indonesia
inertial	motion unaffected by forces after initiation
interglacial	the periods in the Quaternary when global ice cover was a minimum. Occur roughly every 100 000 years. We are currently in an interglacial period
internal	processes occurring at boundaries between regions of different density within a fluid
isopycnal	a surface of constant density in the ocean

jet-stream	a zone of extremely strong winds in the upper *troposphere*
(coastal) Kelvin wave	a travelling disturbance in the upper ocean, confined to coastal regions and always moving with the coast on the right (in the Northern Hemisphere). Its peak amplitude is at the coast itself
laminar	layered or regular
latent heat	the energy released by water vapour upon condensation
luminescence	radiation of energy by an electron in an atom or molecule as it returns to its standard electron configuration
mixed layer	the upper region of the ocean well-mixed by interaction with the overlying atmosphere
momentum transfer	the inducing of motion in another body by physical means such as collision
nitracline	the zone in the ocean where the nutrient (NO_3^-) concentration begins to substantially increase above its near-surface value
orbital	an electron energy state within an atom
osmosis	the transfer of gases and liquids across cell walls
over-turning	the process induced in the ocean or atmosphere when a heavier fluid over-lies a lighter
oxidation state	number of electrons associated with an atom less than those found in the elemental state
ozone layer	region of the stratosphere with high ozone concentrations
pack ice	extensive, free-floating ice
parameterisation	approximating a process in a General Circulation Model which has no simple mathematical description
partial derivative	a rate of change of a variable in one direction only
partial pressure	the pressure exerted by one gas alone in a mixture
perihelion	the point of the Earth's orbit closest to the Sun
permanent thermocline	the region of rapid vertical temperature change at the deepest extent of winter mixing in the ocean
phosphorescence	similar to fluorescence but with energy emission occurring after the initiating irradiation has ceased
photo-dissociation	breakdown of a molecule by radiation
photo-ionization	release of an electron from an atom by radiation
photolytic	a reaction that depends on *electromagnetic radiation* of a particular wavelength or wavelength band
photon	a discrete particle of *electromagnetic radiation*
photosphere	layer in solar atmosphere from which light and heat originate
photosynthesis	conversion of solar radiation into sugar by plants
phytoplankton	microscopic plants in the upper ocean
planetary boundary layer	the region of the atmosphere bordering the ground or sea surface which is actively involved in the exchange of gases, heat and momentum between the surface and the atmosphere as a whole
planetary vorticity	*vorticity* due to the underlying rotation of the Earth
plankton	any living organism within the sea restricted to movement with the prevailing current
plate tectonics	the theory that the Earth's crust is composed of a set of discrete plates that move over its surface due to motion imposed from deeper within the planet

potential temperature	the temperature that a parcel of air or water would have if it was taken *adiabatically* to sea level
primary productivity	rate of production of *plankton*
proxy	a measure of some quantity that is indirectly linked to the variable of interest
pseudo-force	a force needed to explain processes seen within a moving coordinate system but absent from the perspective of an external, stationary, reference frame. Examples are Coriolis and centrifugal forces
Quasi-Biennial Oscillation (QBO)	zonally symmetric oscillation in the circulation of the lower stratosphere occurring with a roughly two year period
radical	a charged atom or molecule
Rayleigh scattering	scattering by particles smaller than the wavelength of the scattered wave
reduction	the opposite of oxidation; the loss of an oxygen atom or its equivalent
reflectance	the proportion of radiation reflected directly from a surface. This is not the same as albedo
refractive index	the ratio of the speed of light in a vacuum to that in the medium of interest
relative vorticity	*vorticity* due to the rotational motion of an object over the surface of the Earth
residence time	the average time – milliseconds to millions of years – that a substance remains in a given reservoir of its cycle
rotational energy	the energy absorbed by a molecule that causes it to rotate
roughness length	the height above the surface at which the flow stops due to friction
saturation	the atmospheric state where no more water vapour can be evaporated without condensation occurring. In a liquid, the state where no more substance (e.g. salt) can be dissolved without the formation of a suspension of crystals
scattering	the phenomenon by which light is forced to deviate from a straight path by interaction with molecules or particles in its path. The light does not, however, pass through, or chemically react with, the obstructing material
seasonal thermocline	the zone of rapid vertical temperature change at the base of the summer ocean mixed layer
sea state	the character of the waves on the sea surface – depends on the wind speed
sensible heat	heat energy transferred by conduction
shear	a vertical gradient of some property
shelf ice	ice in the sea, but ultimately attached to a land glacier
sigma	the ratio of local atmospheric pressure to that at the surface directly beneath
solar constant	the energy flux from the Sun reaching the Earth's orbit
Southern Oscillation specie	the quasi-periodic oscillation in the trans-Pacific pressure gradient between Indonesia and the southeast Pacific a molecule of a given type
specific humidity	the mass of water vapour in a kilogram of dry air

spectral	representation of a quantity in terms of waves
spectrum	the distribution of energy per wavelength with which a given object emits electromagnetic radiation
spring bloom	the rapid growth of plankton in the ocean typically observed in the spring with the return of warmth and plentiful solar radiation
Stefan–Boltzmann Law	dependence of total energy flux emitted by a black body on temperature. *See* equation (1.1)
stratosphere	layer of atmosphere above troposphere, principally important because of the abundance of ozone. *See* Fig. 1.3
sublimation	the exchange of molecules between the vapour and solid state, particularly over ice
sunspot	a dark region in the Sun's photosphere of higher temperature
superposition	the spatial overlapping of two influences
swell	the background wave field caused by disturbances to the sea surface remote from the observation site
teleconnection	near simultaneous climatic link between two or more regions, usually deriving from climatic changes in a base region
thermal	a parcel of warm air rising with respect to its surroundings
thermohaline circulation	the very large-scale circulation of the ocean, driven by density differences
trace gas	a gas which contributes less than about 1% of the total volume of the atmosphere
transfer velocity	the rate at which a gas is exchanged between the air and sea
transient response	climatic change arising during change in one of the climate system's constituents
transmission	the passage of energy through a medium
tritium	an isotope of hydrogen with two neutrons in the nucleus, as well as a proton
troposphere	bottom layer of atmosphere in which most weather occurs. *See* Fig. 1.3
trough	a region of low pressure without a clearly defined centre
vapour pressure	the proportion of the atmospheric pressure exerted by the water vapour molecules in the column of air overhead
ventilation	exposure of a parcel of sea water to the atmosphere
vibrational energy	the energy absorbed by a molecule that causes its shape to flex
virga	rain which fails to reach the surface
vorticity	a measure of rotation, or spin, analogous to angular momentum
Walker circulation	the atmospheric circulation cell of the equatorial Pacific. *See* Fig. 5.26
warm pool	the region of very warm upper ocean in the west equatorial Pacific (*see* Figs. 2.34 and 5.22)
water mass	a body of water with similar temperature and salinity characteristics gained by exposure to the atmosphere in a particular geographic area

wave number	gives the wavelength of upper tropospheric circulation fields. A wave number n has a wavelength of $360/n$ degrees of longitude
wave packet	a discrete, spatially limited, region of wave energy
zooplankton	microscopic animals in the *plankton*

Bibliography

Literature dealing with the oceans' role in climate is vast, and growing exponentially in volume. The scope of this book makes it impractical to reference all works consulted during its writing. Such a list would be out of date before the book was published. However, the reader new to the field needs some direction about where to go next. Thus, at the end of each chapter key books or articles are listed, with a short description, that provide additional avenues of exploration. This bibliography gives the full details of these texts, and also for other references, such as those used for diagrams or from which data were obtained.

Barron, E. J., J. L. Sloan II, and C. G. A. Harrison (1980): Potential significance of land–sea distribution and surface albedo variations as a climatic forcing factor; 180 My to the present. *Palaeogeogr., Palaeoclimatol., Palaeoecol.*, **30**, 17–40.

Barron, E. J. and W. M. Washington (1984): The role of geographic variables in explaining paleoclimates:results from Cretaceous climate model sensitivity studies. *J. Geophys. Res.*, **89D**, 1267–79.

Baumgartner, A. and E. Reichel, (1975): *The world water balance: mean annual global, continental and marine precipitation and run-off.* Elsevier (Amsterdam), 179 pp.

Berger, A. L. (1978): Long-term variations of caloric insolation resulting from the Earth's orbital elements. *Quaternary Res.*, **9**, 139–67.

Bigg, G. R., and J. R. Blundell (1989): The equatorial Pacific Ocean prior to and during El Niño of 1982/83 – a normal mode view. *Quart. J. Roy. Meteorol. Soc.*, **115**, 1039–69.

Bigg, G. R. (1990): El Niño and the Southern Oscillation, *Weather*, **45**, 2–8.

Bigg, G. R. (1992a): The atmospheric engine. *Geographical*, **64(1)**, Analysis, 1–3.

Bigg, G. R. (1992b): Droplets to thunderstorms. *Geographical*, **64(2)**, Analysis, 5–7.

Bigg, G. R. (1992c): Climates of the past. *Geographical*, **64(5)**, 52–3.

Bigg, G. R. (1992d): Climates of the future. *Geographical*, **64(6)**, 50–2.

Bigg, G. R. (1993): Comparison of coastal wind and pressure trends over the tropical Atlantic: 1946–87. *Int. J. Climatol.*, **13**, 411–21.

Bigg, G. R. (1994): Beyond the weather map. *Geographical*, **66(10)**, 55–6.

Bond, G., H. Heinrich, W. Broecker, L. Labeyrie, J. McManus, J. Andrews, S. Huon, R. Jantschik, S. Clasen, C. Simet, K. Tedesco, M. Klas, G. Bonani, and S. Ivy (1992): Evidence for massive discharges of icebergs into the North Atlantic ocean during the last glacial period. *Nature*, **360**, 245–360.

Bradley, R. S., and P. D. Jones (1993): Records of explosive volcanic eruptions over the last 500 years. In *Climate since A.D. 1500*, eds. R. S. Bradley and P. D. Jones, Routledge (London), pp. 606–22.

Broecker, W. S., and T.-H. Peng (1982): *Tracers in the sea.* Eldigio Press (Palisades, New York), 690pp.

Broecker, W. S., and T.-H. Peng (1992): Interhemispheric transport of carbon dioxide by ocean circulation. *Nature*, **356**, 587–9.

Broecker, W. S., M. Andree, W. Wolfli, H. Oeschger, G. Bonani, J. Kennett, and D. Peteet (1988): The chronology of the last deglaciation: implications to the cause of the Younger Dryas event. *Paleoceanogr.*, **3**, 1–19.

Chahine, M. T. (1992): The hydrological cycle and its influence on climate. *Nature*, **359**, 373–9.

Clemens, S., W. Prell, D. Murray, G. Shimmield, and G. Weedon (1991): Forcing mechanisms of the Indian Ocean monsoon. *Nature*, **353**, 720–5.

Climate Diagnostics Bulletin (1994a). No. **94/1**, National Meteorological Center (Washington), 77 pp.

Climate Diagnostics Bulletin (1994b). No. **94/2**, National Meteorological Center (Washington), 79 pp.

Climate Diagnostics Bulletin (1994c). No. **94/5**, National Meteorological Center (Washington), 83 pp.

Climate Diagnostics Bulletin (1995). No. **95/3**, National Meteorological Center (Washington), 83 pp.

Cooper, N. S., G. R. Bigg, and K. D. B. Whysall (1989): Recent decadal climate variation in the tropical Pacific. *Int. J. Climatol.*, **9**, 221–42.

Crowley, T. J., and G. R. North (1991): *Paleoclimatology*. Oxford University Press (Oxford), 339 pp.

Davies, A. M., (1986): Mathematical formulation of a spectral tidal model. In *Advanced physical oceanographic numerical modelling*, ed. J. J. O'Brien, Reidel (Dordrecht), pp. 373–90.

Fairbanks, R. G. (1989): A 17,000 year glacio-eustatic sea level record: Influence of glacial melting rates on the Younger Dryas event and deep-ocean circulation. *Nature*, **342**, 637–42.

Frakes, L. A. (1979): *Climates through geologic time*, Elsevier (Amsterdam), 310 pp.

Frakes, L. A., J. E. Francis, and J. I. Syktus (1992): *Climate Modes of the Phanerozoic*, Cambridge University Press (Cambridge), 274pp.

Gill, A. E. (1982): *Atmosphere–ocean dynamics*. Academic Press (London), 662pp.

Goudie, A. (1992): *Environmental Change*. 3rd edition, Clarendon (Oxford), 329 pp.

Graham, N. E., and W. B. White (1988): The El Niño cycle: a natural oscillator of the Pacific Ocean–atmosphere system. *Science*, **240**, 1293–1302.

Han, Y.-J. (1988): Modelling and simulation of the general circulation of the ocean. In *Physically-based modelling and simulation of climate and climatic change*, ed. M. E. Schlesinger, Kluwer (Dordrecht), pp. 465–508.

Haq, B. U., and F. W.B. Van Eysinga (1987): *Geological time chart*. Elsevier (Amsterdam).

Hartmann, D. L. (1994): *Global physical climatology*. Academic Press (San Diego), 411pp.

Harvey, J. G. (1985): *Atmosphere and Ocean: our fluid environment*. Artemis (London), 143pp.

Hide, R., and J. O. Dickey (1991): Earth's variable rotation. *Science*, **253**, 629–37.

Horel, J. D., and J. M. Wallace (1981): Planetary scale atmospheric phenomena associated with the interannual variability of sea-surface temperature in the equatorial Pacific. *Mon. Wea. Rev.*, **109**, 813–29.

Houghton, J. T., G. J. Jenkins, and J. J. Ephraums (eds.) (1990): *Climate Change – the IPCC scientific assessment*. Cambridge University Press (Cambridge), 365pp.

Houghton, J. T., B. A. Callander, and S. K. Varney (eds.) (1992): *Climate Change 1992: the supplementary report to the IPCC Scientific Assessment*. Cambridge University Press (Cambridge), 200pp.

Houghton, J. T., L. G. Meiro Filho, J. Bruce, H. Lee, B. A. Callander, E. Haites, N.

Harris, and K. Maskell (eds.) (1995): *Climate Change 1994: Radiative forcing of climate change and an evaluation of the IPCC IS92 emission scenarios.* Cambridge University Press (Cambridge), 339pp.

Howell, D. G. (1993): *Tectonics of suspect terranes.* Chapman and Hall (London), 232pp.

Huggett, R. J. (1991): *Climate, earth processes and earth history.* Springer-Verlag (New York), 280 pp.

Imbrie, J., J. D. Hays, D. G. Martinson, A. McIntyre, A. C. Mix, J. J. Mauley, N. G. Pisias, W. L. Prell, and N. J. Shackleton (1984): The orbital theory of climate – support from a revised chronology of the marine $\delta^{18}O$ record. In *Milankovitch and climate: understanding the response to astronomical forcing*, ed. A. Berger, Reidel (Dordrecht), pp. 269–305.

James, D. A., and I. Simmonds (1993): A climatology of Southern Hemisphere extratropical cyclones. *Climate Dynamics*, **9**, 131–45.

Jenkins, W. J. (1988): The use of anthropogenic tritium and helium-3 to study subtropical gyre ventilation and circulation. *Phil. Trans. R. Soc. Lond. A*, **325**, 43–61.

Jerlov, N. G. (1976): *Optical Oceanography.* Elsevier (Amsterdam), 194pp.

Keeling, C. D. (1968): Carbon dioxide in surface ocean waters 4. global distribution. *J. Geophys. Res.*, **73**, 4543–53.

Kemp, D. D. (1994): *Global Environmental Issues*, 2nd edition. Routledge (London), 224pp.

Lamb, H. H. (1977): *Climates of the past, present and future. volumes I and II* Metheun (London).

Lehman, S. J., and L. D. Keigwin (1992): Sudden changes in North Atlantic circulation during the last deglaciation. *Nature*, **356**, 757–62.

Levitus, S. (1982): *A climatological atlas of the world ocean.* NOAA Prof. Pap. No. 13, U. S. Government Printing Office.

Ludlam, F. H. (1980): *Clouds and storms: the behaviour and effect of water in the atmosphere.* Pennsylvania State University Press (Philadelphia), 405 pp.

Maier-Reimer, E., and U. Mikolajewicz (1990): Ocean general circulation model sensitivity experiment with an open central American isthmus *Paleoceanogr.*, **5**, 349–66.

Manahan, S. E. (1990): *Environmental chemistry*, 4th edition. Lewis (New York), 612pp.

Mann, K. H., and J. R. N. Lazier (1991): *Dynamics of marine ecosystems.* Blackwell (Oxford), 466pp.

Mason, B. J. (1971): *The physics of clouds*, 2nd edition. Clarendon Press (Oxford), 671pp.

McIlveen, R. (1992): *Fundamentals of weather and climate.* Chapman and Hall (London), 497pp.

Mitchell, J. F. B., N. S. Grahame, and K. J. Needham (1988): Climate simulation for 9000 years before present: seasonal variations and the effect of the Laurentide ice sheet. *J. Geophys. Res.*, **94**, 16097–114.

Nobre, C. A., P. J. Sellers, and J. Shukla (1991): Amazonian deforestation and climate change. *J. Climate*, **4**, 957–88.

Oberhuber, J. M. (1988): *Atlas of COADS data.* Max-Planck Institute (Hamburg).

Oort, A. H. (1983): *Global atmospheric circulation statistics 1958–73.* NOAA Prof. Paper No. **14**, U. S. Govt. Printing Office, Washington D. C.

Oort, A. H. (1988): Climate observations and diagnostics. In *Physically-based modelling and simulation of climate and climatic change*, ed. M. E. Schlesinger, Kluwer (Dordrecht), 813–40.

Open University Course Team (1989) *Ocean Circulation.* Pergamon Press (Oxford), 238pp. Other titles in chemistry, waves and sediments also exist.

Owens, N. J. P., C. S. Law, R. F. C. Mantoura, P. H. Burkill and C. A. Llewellyn, *Nature,* **354**, 293–5.

Parrish, J. M., J. T. Parrish, and A. M. Ziegler (1986): Permian–Tertiary paleogeography and paleoclimatology and implications for therapsid distribution. In *The ecology and biology of mammal-like reptiles,* eds. N. Hotton II, P. D. MacLean, J. J. Roth, and E. C. Roth, Smithsonian Institution Press (Washington), 109–31.

Parsons, T. R., M. Takahashi, and B. Hargrave (1984): *Biological oceanographic processes.* Pergamon (Oxford), 330 pp.

Peltier, W. R. (1994): Ice age paleotopography. *Science,* **265**, 195–201.

Philander, S. G., T. Yamagata, and R. C. Pacanowski (1984): Unstable air-sea interactions in the tropics. *J. Atmos. Sci.,* **41**, 604–13.

Philander, S. G. (1990): *El Niño, La Niña, and the Southern Oscillation.* Academic Press (San Diego), 293pp.

Pickard, G. L., and W. J. Emery (1982) *Descriptive physical oceanography.* Pergamon Press (Oxford), 249pp.

Pinet, P. R. (1992): *Oceanography.* West (St. Paul), 571pp.

Pond, S., and G. L. Pickard (1983): *Introductory dynamical oceanography.* Pergamon Press (Oxford), 329pp.

Prentice, I. C., W. Cramer, S. P. Harrison, R. Lehmans, R. A. Manserud, and A. M. Solomon (1992): A global biome model based on plant physiology and dominance, soil properties and climate. *J. Biogeogr.,* **19**, 117–34.

Quinn, W. H., and V. T. Neal (1993): The historical record of El Niño events. In *Climate since A.D. 1500,* eds. R. S. Bradley and P. D. Jones, Routledge (London), pp. 623–48.

Raiswell, R. W., P. Brimblecombe, D. L. Dent, and P. S. Liss (1980): *Environmental chemistry.* Edward Arnold (London), 184pp.

Rasmussen, E. M., and T. H. Carpenter (1982): Variation in tropical sea surface temperature and surface wind fields associated with the Southern Oscillation/El Niño. *Mon. Wea. Rev.,* **110**, 354–84.

Raymo, M. E., W. F. Ruddiman, N. J. Shackleton, and D. W. Oppo (1990): δ^{13}C gradients over the last 2.5 million years. *Earth Planet. Sci. Lett.,* **97**, 353–68.

Ropelewski, C. F., and M. S. Halpert (1987): Global and regional scale precipitation associated with El Niño/Southern Oscillation. *Mon. Wea. Rev.,* **115**, 1606–26.

Rogers, R. R., and M. K. Yau (1991): *A short course in cloud physics,* 3rd edition. Pergamon Press (Oxford), 293pp.

Rostek, F., G. Ruhland, F. C. Bassinot, P. J. Mller, L. D. Labeyrie, Y. Lancelot, and E. Bard (1993): Reconstructing sea surface temperature and salinity using δ^{18}O and alkenone records. *Nature,* **364**, 319–21.

Schwartz, S. E. (1988): Are global cloud albedo and climate controlled by marine phytoplankton? *Nature,* **336**, 441–5.

Street-Perrott, F. A., and S. P. Harrison (1985): Lake levels and climate reconstruction. In *Paleoclimate analysis and modeling,* ed. A. D. Hecht, Wiley (Chichester), 291–340.

Taljaard, J. J., and H. Van Loon (1984): Climate of the Indian Ocean south of 35°S. In *World survey of climatology, volume 15, climates of the oceans,* ed. H. Van Loon, Elsevier (Amsterdam), pp.505–601.

Terada, K., and M. Hanzawa (1984): Climate of the North Pacific Ocean. In *World survey of climatology, volume 15, climates of the oceans,* ed. H. Van Loon, Elsevier (Amsterdam), pp. 431–503.

Thunell, R. C., D. F. Willams, and M. B. Cita (1983): Glacial anoxia in the eastern

Mediterranean. *J. Foram. Res.*, **13**, 283–90.

Veizer, J., P. Fritz, and B. Jones (1986): Geochemistry of brachiopods: oxygen and carbon isotopic records of Paleozoic oceans. *Geochimica et Cosmochimica Acta*, **50**, 1679–96.

Veum, T., E. Jansen, M. Arnold, I. Beyer, and J.-C. Duplessey (1992): Water mass exchange between the North Atlantic and the Norwegian Sea during the past 28,000 years. *Nature*, **356**, 783–5.

Wallace, J. M., and D. S. Gutzler (1981): Teleconnections in the geopotential height field during the Northern Hemisphere winter. *Mon. Weath. Rev.*, **109**, 784–812.

Webster, P. J. (1983): Large-scale structure of the tropical atmosphere. In *Large-scale dynamical processes in the atmosphere*, eds. B. J. Hoskins and R. P. Pearce, Academic Press (New York), 235–75.

Webster, P. J., and S. Yang (1992): Monsoon and ENSO – selectively interactive systems. *Quart. J. Roy. Meteorol. Soc.*, **118**, 877–926.

Woods, J. D. (1984): *The warmwatersphere of the North East Atlantic. A miscellany.* Inst. Meereskunde Kiel, Report 128 39 pp.

Index